Digital Control

Digital Control
Fundamentals, Theory and Practice

W. Forsythe and R. M. Goodall

Department of Electronic & Electrical Engineering
Loughborough University of Technology

McGraw-Hill, Inc.

New York St. Louis San Francisco Bogotá Caracas
Mexico Montreal San Juan São Paulo Toronto

1 2 3 4 5 6 7 8 9 0 DOC/DOC 9 7 6 5 4 3 2 1

ISBN 0-07-021600-2

First published 1991 by MACMILLAN EDUCATION LTD.
Houndmills, Basingstoke, Hampshire RG21 2XS
and London.

Contents

Preface ix

1 Introduction **1**

2 The z-Domain **4**

2.1 Continuous and discrete representation of a mathematical function of time 4

2.1.1 Tustin's method and the trapezoidal rule 5

2.2 A polynomial transformation 8

2.3 The discrete transfer function 11

2.4 The link between s and z 13

2.5 The z-transform table 14

2.6 Use of the table 17

2.6.1 The impulse 18

2.6.2 Mathematical relationships within the loop 19

2.7 Why use the transform table? 24

2.8 Points of nomenclature 24

2.9 Including time delay 25

2.10 The impulse sampler and the star notation 28

2.11 Analysis of sampled data systems 29

Problems 38

3 Controller Design **44**

3.1 The frequency response of a digital filter 45

3.1.1 Aliasing 45

3.1.2 Zero frequency response 48

3.1.3 Initial and Final Value Theorems 50

3.2 Emulation 51

3.2.1 The bilinear transformation 51

3.2.2 Adams–Bashforth algorithms 54

3.2.3 Use of the derivative 55

3.2.4 Pole and zero mapping 56

3.2.5 Bilinear transform with pre-warping 57

3.2.6 A pole mapping technique 59

3.2.7 Comparisons 59
3.3 Controller design on the w-plane 61
3.4 Some properties of the z-plane 69
3.5 The frequency response of sampled data systems 72
3.6 Compensation of a complex sampled data system 75
3.7 Multiple sampling rates 79
3.8 Summary 80
Problems 81

4 Error Mechanisms, Filter Structure, and the Sampling Interval 88
4.1 Errors in digital filters 89
4.1.1 Algorithmic error 89
4.1.2 Quantisation error 90
4.2 The significance of structure 92
4.3 Variation of filter error with sampling interval 95
4.3.1 e_a: algorithmic error 96
4.3.2 e_s: sampler quantisation error 97
4.3.3 e_r: multiple word truncation error 99
4.3.4 e_c: coefficient representational error 104
4.3.5 Summary 106
4.4 Sampling interval and loop stability 106
4.4.1 Gauging the reduction in stability 107
4.5 Rationale for the choice of system parameters 108
4.6 Numerical evaluation of e_a, e_c, e_s 109
4.6.1 e_a, the algorithmic error 109
4.6.2 e_c, the coefficient representational error 110
4.6.3 e_s, the sampler quantisation error 110
4.7 Conclusions 111

5 Requirements for Implementation 112
5.1 Analogue *vs* digital controllers 112
5.2 Generalised architecture of a digital controller 115
5.3 Functional representation 117
5.4 Filter structures 120
5.4.1 The direct form 120
5.4.2 The canonic form 121
5.5 Determining the precision of computation 123
5.6 Coefficient wordlength 124
5.6.1 General procedure for choice of coefficient wordlength 125
5.6.2 Sensitivity analysis of a first-order filter 125
5.6.3 Sensitivity analysis of a second-order filter 131
5.6.4 Second-order filters which can be factorised 136

5.6.5 Summary of coefficient accuracy requirements 137
5.7 The internal variables 138
5.7.1 Internal variable overflow 138
5.7.2 Overflow in general 141
5.7.3 Internal variable underflow 143
5.8 Dealing with integrators 146
5.8.1 Discrete transfer functions with integrators 146
5.8.2 Coefficient wordlength 147
5.8.3 Overflow of the internal variable 148
5.9 Implementation using the δ-operator 151
5.9.1 The principle of the δ-operator 152
5.9.2 Structures of the δ-filter 153
5.9.3 Coefficient wordlength 157
5.9.4 Internal variable overflow 161
5.9.5 Internal variable underflow 163
5.9.6 Integrators using δ 165
5.10 Summary of the effects of digitisation 168

6 Specific Issues of Implementation 171
6.1 Hardware for digital controllers 171
6.1.1 Types of processor 173
6.1.2 Timer/counters 177
6.1.3 Input/output devices 178
6.1.4 Development and testing 181
6.1.5 Example control circuits 183
6.2 Software considerations 187
6.2.1 Software structure 187
6.2.2 Numerical routines 198
6.2.3 Languages 204

Appendix A: Table of Laplace and z-transforms 216

Appendix B: Block diagram representation of a transfer function 218

Appendix C: A mathematical description of the sampling process 220

Appendix D: Compensation techniques 224

Appendix E: Emulation using pole mapping, with zero positions determined using the Taylor expansion 234

Appendix F: Analysis of the phase advance compensator 242

Appendix G: A second-order filter 248

Appendix H: Design Structure Diagrams – a brief guide 249

References 251

Index 254

Preface

The purpose of this text is to examine the problems inherent in the use of a digital processor for purposes of control, problems both of theory and of practice. Many volumes have already been written with this aim but in our opinion there is room for yet another; there are things that need to be said, important issues that need to be clarified, which current textbooks tend to ignore or to deal with inadequately in one respect or another. The philosophy is not to offer the reader everything we know, but to restrict the information to little more than the essentials. This has two desirable effects: the size and therefore the cost of the book is kept to a minimum and, secondly, the beginner is not called upon to distinguish in his studies between what is and what is not *necessary* for a proper understanding of the subject.

Throughout the book we have attempted to keep the needs of the beginner in mind; the z-transform, for example, is in our experience less understood than its continuous counterpart, the Laplace or s-transform. This cannot be explained purely in terms of the mathematics since the z-transform is the simpler concept. The problem lies, we believe, in the conventional exposition, so the attempt has been made to develop the z-transform on an intuitive basis, to show that the mathematical definition 'makes sense'. The modified z-transform is dealt with too, since this requires only a slight extension of the idea.

Some trouble is taken to explain the operation of the sampled data loop and the steps needed to construct a suitable mathematical description of it. This entails clarification of such basic concepts as the impulse and an explanation of the meaning and use of the z-transform table. Perhaps surprisingly, it appears to be one of the chief complaints of the beginner that it is difficult to gain a clear perception of these very basic notions.

Methods of designing digital compensation for the sampled data loop are presented, and a number of them discussed in some detail – those that in our estimation the reader will most readily pick up and which will serve for most occasions.

One of the most difficult topics to write meaningfully about is the choice of sample interval, yet it cannot be avoided. It is treated here at some length and is discussed in the context of error generation and filter structure, to both of which it is intimately related.

Many texts attempt to deal with the basics of digital control without giving much attention to the practical issues of implementation: yet there are a number of difficulties that arise and pitfalls that must be avoided when implementing even quite simple controllers. Considerable space is therefore devoted to examination of the fundamental issues that need to be considered when implementing a digital controller: quantisation of the coefficients and internal variables, the implications of this with respect to computational requirements, the hardware and software aspects of the controller itself. In a number of cases both equations and software structures are illustrated by means of program code, and we have chosen to use C because of its applicability to real time applications. We have restricted such coding to the most basic program constructions, and the reader should be able to interpret them without prior knowledge of the language.

The problems of realisation that are dealt with here use examples which are typically encountered in the design of electromechanical servo systems rather than industrial process control. The same realisation problems occur in both types of application [Be88], but they are generally more acute in the former because of the higher speed of operation. The theoretical concepts that are explained underlie control systems of all types.

As well as emphasising what is to be found in the pages that follow, it is necessary to say something about what is not. The most notable omission is perhaps an account of the mathematics of the z-operator. The exclusion of this material is in line with the philosophical approach mentioned above; one can work quite satisfactorily in discrete terms without a knowledge of the more arcane features of the z-domain, provided, that is, the z-transform table is taken as given.

Since it makes no sense to embark on a study of digital control without previous acquaintance with at least the basic ideas of classical control theory, the assumption is made that the reader understands the rudiments of that subject. Given that this is the case, and not very much is presumed, the reader should have little difficulty in grasping the few theoretical concepts that are presented.

Digital control turns out to be computationally very intensive so that the execution of even the simplest ideas entails a great deal of tedious calculation. It helps very much to have available some software package that will largely remove the burden of the computation, and we are willing to supply a copy of our own suite of programs to anyone who wants it. Those interested should write to the address below for details of price (which is very modest) and availability. The software in question performs transformations between the s- and z-domains via the transform table (including the table of modified transforms), the bilinear transformation, and the pole mapping transform described in Appendix E of this book. It is also very desirable to have available some control system design software to carry out the exercises suggested, or to validate examples quoted; quite a few such packages can now be bought. Many of the plots to be found in these pages were drawn using Simbol 2, from Cambridge Control Ltd. Those readers engaged in teaching, either as teachers or students, might be interested in the *Manual of Solutions* which is again available on request, at the address below.

Finally, we must acknowledge our great debt to Jean Grantham and Christine Sharpe for their invaluable assistance with the production of the present volume. Their good humour and infinite patience are much appreciated.

Department of Electronic and Electrical Engineering,
Loughborough University of Technology,
Loughborough,
Leicestershire LE11 3TU

Bill Forsythe
Roger Goodall

1 Introduction

Over the last thirty years or so, control has become a well recognised branch of engineering, establishing itself over an enormous range of disciplines and applications. That range increased still further with the advent of the microprocessor; inexpensive 'intelligence' became suddenly and extensively available. Unfortunately digital techniques have not proved easy to apply. Those very features that make the microprocessor so attractive also make it difficult to employ; there are a greater number of important decisions to be made when designing a digital controller than when designing an analogue one.

It is also the case that engineers as a group are much less familiar with the z-domain algebra used for discrete systems than with the s-domain algebra used for continuous systems. For some people at least, this constitutes a serious handicap; one naturally feels ill at ease dealing with concepts that are only partially understood.

It is not that the basic notions of control have altered in some fundamental way with the introduction of digital components; anyone with a 'feel' for control systems will usually find that his intuition still leads him to good solutions. His task however is more difficult, and when he turns to mathematics to confirm his intentions or to fine-tune his solution, he may not be able to use it properly. One purpose of this book is therefore to acquaint the engineer or the student with the thinking behind the mathematics of the z-transform and its application in control systems; one of the impediments that lie between the engineer and the achievement of his aims should thus be removed.

So why use a digital controller? Before grappling with the problems of implementing one, it is worth reflecting upon the motivation for its use; in particular it is worth comparing the digital with its analogue counterpart.

The 'processors' of analogue controllers are operational amplifiers. The mathematical functions which they implement are determined by the passive components which are associated with them. The combination gives virtually infinite-resolution delay-free processing at very modest cost. Extra functions, such as will be required for multiloop controllers, are provided by additional operational amplifier stages, facilitated by the availability of up to four operational amplifiers in a single package. Each of the signals which link the various functions within a complete analogue controller is carried by a single track on the printed circuit board. It is a straightforward matter to design general-purpose compensation stages which can be adapted to meet a variety of requirements by means of links, adjustable or 'select on test' resistors, etc. Making changes requires the sophistication of a soldering iron and sidecutters.

The processors of digital controllers may be microprocessors, single chip microcontrollers, signal processors or even speciai-purpose processors. All of these need A/D and D/A converters for most applications, and they are likely to need some external memory, in particular program memory containing the software which determines the mathematical functions they implement. Their signals require multitrack layouts on the printed circuit board (carrying say eight-bit bytes in parallel). The ability to change functions by software rather than hardware alterations is attractive, but it necessitates access to a microprocessor development system (MDS).

The intention in this comparison is not to paint a negative picture of digital control but rather to ensure that the implications are seen properly and in perspective, 'warts and all'. The combination of sample delays, computation delays and quantisation effects almost invariably means that a digital controller will ultimately give a performance inferior to that of an analogue one, should the latter be a feasible proposition in the circumstances. It is important to realise that the design of an effective digital controller, particularly when compared with that of an equivalent analogue controller, is not a trivial process.

However, there are at least three reasons for resorting to digital control:

(1) *Cost* – whereas an analogue controller requires a number of processing elements to carry out the various functions, a single digital controller can be programmed to carry out all the tasks in sequence. There is therefore the potential for lower cost, particularly with the more complicated controllers in lower-speed systems. The cheapness of the basic analogue processing element is however a significant counter-argument.
(2) *Adaptive control* – if in a particular application a comparable or adequate performance can be achieved with a digital system, *and* if there is computational capacity to spare, the possibility exists for adaptive control in which the 'spare capacity' can be used to monitor the system and adapt the control parameters and/or structure in order to maintain optimal performance [As89]. Adaptive control is something which, while theoretically feasible, is practically very difficult to implement in analogue components, and if the application demands it the adaptability of digital controllers may override other considerations.
(3) *Volume production* – in some circumstances with large quantity applications of a particular product involving a control system, integration

onto full- or semi-custom silicon is likely to be more straightforward with digital components, although the analogue interface requirements must be remembered. Nevertheless the single-chip solutions that may be possible using digital control can be commerically very attractive.

Thus another purpose in what follows is to identify the crucial issues in implementing digital controllers as efficiently as possible, taking into account the fundamental algorithms, the hardware and the software. The emphasis will be upon implementing recursive digital filters to give the usual types of compensator – phase advance, notch filter, etc. The implementation of adaptive strategies however is beyond the scope of a text dealing with the fundamentals of digital control.

It is the authors' belief that frequency response techniques are in some way more fundamental to the design of servo systems than the more recently developed state variable methods, though of course it is best to understand both. As a consequence of this view, the book is devoted to the transfer function approach.

In the matter of nomenclature, a few terms need to be explained at the outset:

- *continuous* refers to the mathematical description and implies the use of s or t
- *discrete* is also a reference to the mathematical description and normally means that time advances in discrete steps; note that one may have a discrete description of a continuous system (as explained later)
- *analogue* refers to the technology involved, the hardware in which the system or component is constructed
- *digital* is also a reference to the hardware; note that both time and amplitude are quantised.

In the pages that follow, the authors present the ideas, the information and the techniques that they regard as either desirable or essential for anyone wishing to understand and especially to build a digital controller.

2 The z-Domain

It is natural for us all to think and write in time domain quantities, but it does not follow that this is the best thing to do in creating a mathematical description of the system we are dealing with.

In the realm of continous systems, it is long established that one gains a great deal from the use of the Laplace or s-transform and much the same advantage is to be had from the z-transform when working with discrete systems. It is necessary however not to draw the analogy between s and z too far; the differences between the two are as great as the similarities, and many of the errors people make in handling z-transforms occur because they forget or do not realise this.

In this chapter therefore:

(1) It is demonstrated using a simple example that in changing from the natural continuous description of a system to a discrete one, we forsake the differential equation for a difference equation.
(2) The z-transform is introduced as a simple artifice designed to facilitate numerical manipulations.
(3) The discrete transfer function is developed.
(4) The difference equation is presented in three ways, all of which are useful: in the time domain, in the z-domain, and as a directed flow graph or digraph.
(5) The z-transform table is introduced and explained, along with the table of modified transforms.
(6) The idea of the impulse sampler and the use of star notation are presented.
(7) The analysis of systems involving several s- and z-blocks in various topologies is studied and a method of handling the analysis is set out.

2.1 Continuous and discrete representation of a mathematical function of time

It is assumed in all that follows that the purpose of implementing a digital controller is to obtain control of some physical plant that can be adequately described by a set of linear ordinary differential equations, the controller and plant forming what is usually referred to as a sampled data system.

2.1.1 Tustin's method and the trapezoidal rule

Any transfer function can be represented in the form of a block diagram (see Appendix B). For example

$$F(s) = \frac{2}{s + 3} = \frac{x_o(s)}{x_i(s)}$$

is shown in figure 2.1. Note that the same thing can be stated in the time domain as

$$\dot{x}_o(t) + 3x_o(t) = 2x_i(t) \tag{2.1}$$

where $\dot{x}_o \triangleq \mathrm{d}x_o/\mathrm{d}t$.

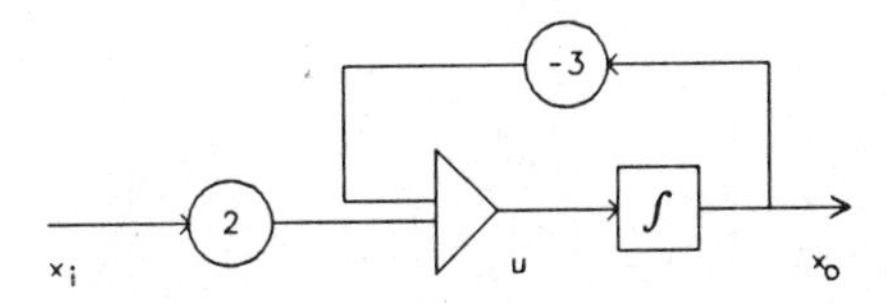

Figure 2.1 Diagrammatic representation of a simple transfer function or differential equation

An obvious way of proceeding to a digital representation of this would be to use a simple digital integration algorithm, and one such is the trapezoidal rule; this avoids complexity yet often gives good results in practice. In figure 2.2 a variable $u(t)$ is depicted, sampled at intervals of T sec. The area under the curve between the latest sample $u(kT)$ and the

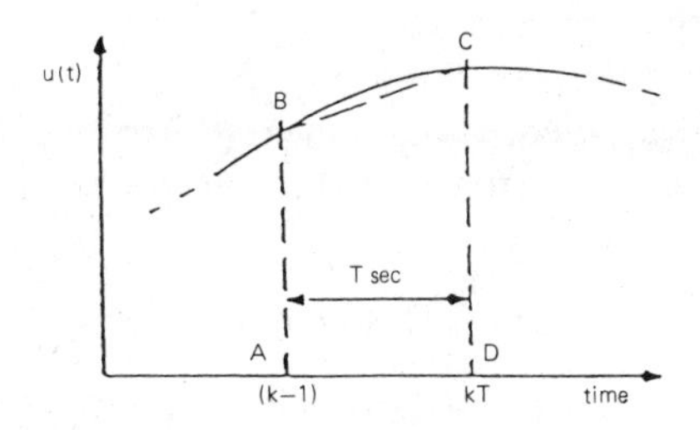

Figure 2.2 Trapezoidal integration

previous sample $u((k-1)T)$ may be approximated by the trapezoid ABCD, the area of which is given by the expression

$$\frac{T}{2}\{u(k) + u(k-1)\}$$

where $u(k) = u(kT)$, $u(k-1) = u((k-1)T)$.

If the integral of $u(t)$ is denoted $x_o(t)$ as in figure 2.1 it can be seen that, provided the rates of change of u between kT and $(k-1)T$ are not great, the trapezoid gives a good approximation to the area under the curve and we may write

$$x_o(k) - x_o(k-1) = \frac{T}{2}[u(k) + u(k-1)] \tag{2.2}$$

From figure 2.1:

$$u(t) = 2x_i(t) - 3x_o(t) \tag{2.3}$$

$$\therefore u(k) = 2x_i(k) - 3x_o(k)$$

and

$$u(k-1) = 2x_i(k-1) - 3x_o(k-1)$$

After substitution in eqn (2.2), this leads to

$$x_o(k) = a_o x_i(k) + a_1 x_i(k-1) - b_1 x_o(k-1) \tag{2.4}$$

where

$$a_o = a_1 = \frac{2T}{2+3T}$$

$$b_1 = -\frac{2-3T}{2+3T}$$

Thus the differential relationship (2.1) has been turned into the difference equation (2.4) using the trapezoidal rule in a manner that is sometimes referred to as Tustin's method.

More complex transfer functions may of course be treated in the same way to yield a difference equation akin to (2.4). Any linear second-order differential equation, for instance, will produce a second-order difference equation of the general form

$$x_o(k) = a_o x_i(k) + a_1 x_i(k-1) + a_2 x_i(k-2) - b_1 x_o(k-1) - b_2 x_o(k-2) \tag{2.5}$$

In a difference equation (such as (2.5)) all the quantities on the right are known either because they are input samples or because they are output samples retained from previous solutions of the equation, so the equation represents an algorithm for the repeated determination of x_o, on the arrival of each new sample of the input.

Difference equations can be presented in diagrammatic form, as a network, just as differential equations can, and figure 2.3 shows one form for eqn (2.5), here referred to as the 'direct form'. Other network structures are possible for the same equation and will be introduced later because they have advantages.

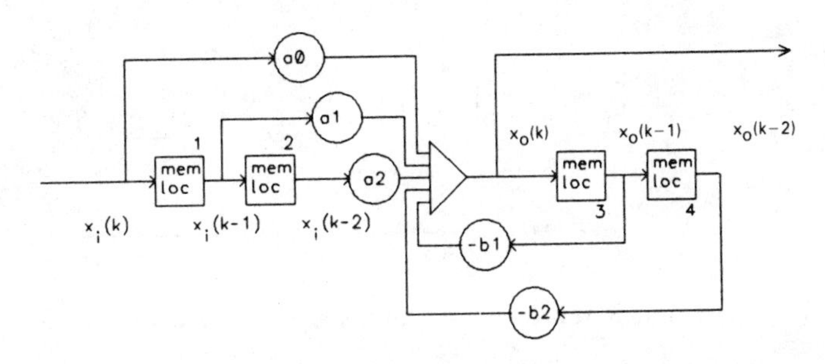

Figure 2.3 Diagrammatic representation of the general second-order difference equation

The diagram may be realised either in hardware or software. The sequence of operations indicated by figure 2.3 in a typical computational cycle is as follows:

- the new sample of the input becomes $x_i(k)$
- the new value of the output, $x_o(k)$, is computed as the diagram indicates – the weighted sum of the various sample values
- the current contents of memory locations are shifted 'down' so that $x_i(k - 1)$ moves to memory location 2 to become $x_i(k - 2)$, $x_i(k)$ moves to memory location 1 to become $x_i(k - 1)$, while $x_o(k)$ becomes $x_o(k - 1)$, etc.

One can assume for the moment that all operations take negligible time so that $x_o(k)$ is available instantly.

It is evident from the simple derivation of eqn (2.4) that the difference equation cannot be an exact transformation of the differential equation because trapezoidal integration is not exact; this is always true no matter how complex the method of integration. There is in general therefore a degree of approximation inherent in the use of digital techniques – the skill lies in keeping the approximation acceptably close.

2.2 A polynomial transformation

The discrete equivalent of the differential equation is the difference equation, as demonstrated above. In this section we look at the task of computing the solution of the difference equation, that is, of determining the output of a discrete network. The network of figure 2.3 would be a general example but it is more convenient to consider a specific example, that of figure 2.4. This has been chosen to make clear a number of general points, as will be seen.

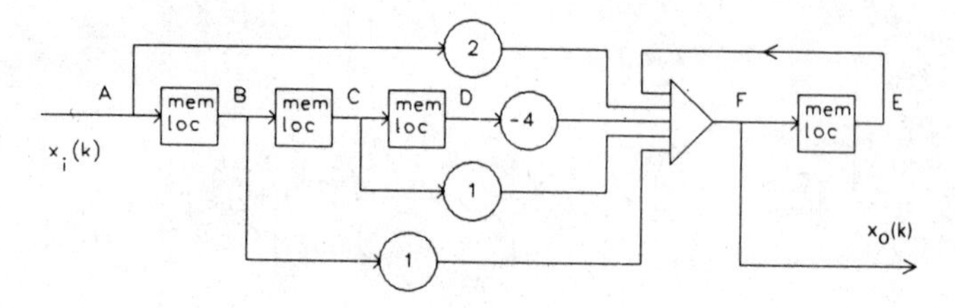

Figure 2.4 Diagrammatic representation of a specific difference equation

Consider a single pulse applied to the input and the resulting sequence of pulses at the output (see figure 2.5). The output is referred to as x_p because it is the response to a single pulse; it can be determined as shown in the time-value table in figure 2.6, given that (see figure 2.4) $F = 2A + B + C - 4D + E$.

Thus when the input is a single pulse of unit amplitude the output is

$$x_p(kT) = \{2, 3, 4\} \tag{2.6}$$

(It is the convention in this text that a sequence of pulses or sample values is represented in braces in the manner of eqn (2.6).)

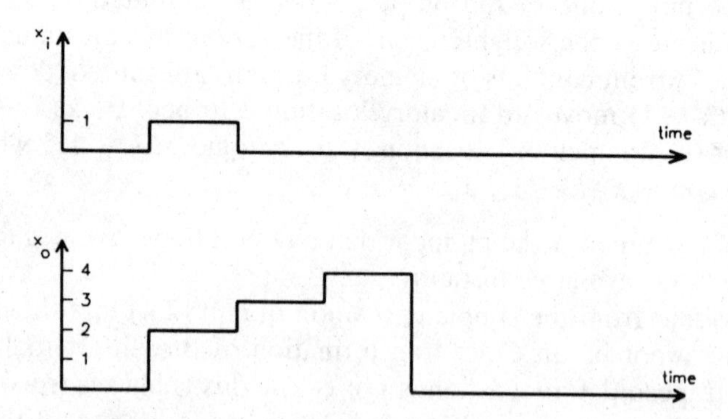

Figure 2.5 The input pulse and resulting output pulses from the system shown in figure 2.4

Time	Value					
	A	B	C	D	E	F
0	1	0	0	0	0	2
T	0	1	0	0	2	3
$2T$	0	0	1	0	3	4
$3T$	0	0	0	1	4	0
$4T$	0	0	0	0	0	0

Figure 2.6 Time-value table for the network of figure 2.4

Consider next a *sequence* of input pulses, say

$$x_i(kT) = \{5, 6, 7\} \tag{2.7}$$

Since everything is linear the responses to pulses of magnitudes 5, 6 and 7 will be respectively

$$5x_p(kT) = \{10, 15, 20\}$$
$$6x_p(kT) = \{12, 18, 24\}$$
$$7x_p(kT) = \{14, 21, 28\}$$

and the total output will be given by the sum of these, allowing for the time difference in the moments of application, that is

$$\begin{array}{l} \{10, 15, 20, 0, 0\} \\ \{0, 12, 18, 24, 0\} \\ \{0, 0, 14, 21, 28\} \\ \hline \end{array}$$

$$x_o(kT) = \{10, 27, 52, 45, 28\} \tag{2.8}$$

There is an obvious similarity between this process, which is called discrete convolution, and the algebraic process of multiplying together two polynomials. For example, construct from the expressions (2.6) and (2.7) the polynomials

$$x_p(v) = 2 + 3v + 4v^2 \tag{2.9}$$

and

$$x_i(v) = 5 + 6v + 7v^2 \tag{2.10}$$

where as yet we attribute no particular significance to v.

Their product gives

$$x_o(v) = 10 + 27v + 52v^2 + 45v^3 + 28v^4 \tag{2.11}$$

Clearly, if eqns (2.9) and (2.10) are derived from eqns (2.6) and (2.7), then (2.8) may be derived from (2.11). Multiplication of the functions in v is thus an alternative to convolution. The transformation of a sequence such as that in eqn (2.6) into a series (2.9) is a simple but effective artifice that transforms convolution in the time domain into multiplication in the v-domain – an advantage when manipulating the numbers.

Three points should be noted:

(1) If (2.11) is multiplied throughout by v it becomes

$$vx_o(v) = 10v + 27v^2 + 52v^3 + 45v^4 + 28v^5 \tag{2.12}$$

and if (2.11) transforms into (or is derived from) (2.8) then (2.12) must transform into the time sequence:

$$x_1(kT) = \{0, 10, 27, 52, 45, 28\}$$

By comparison with (2.8) this is seen to be the same sequence delayed by one sampling interval, so multiplication of the polynomial (2.11) by v corresponds to delay of the pulse sequence (2.8) by one interval. Thus v is found to have a physical significance.

(2) It is obvious that the transformation into the v-domain is only applicable to sequences of numbers and not to a continuous waveform, even though the numbers being dealt with are usually derived from some continous function of time.

(3) A continous function sampled at intervals of T sec to give the infinite sequence

$$x(kT) = \{x(0), x(T), \ldots, x(nT), \ldots\}$$

transforms into the infinite series

$$x(v) = x(0) + x(T)v + \ldots + x(nT)v^n + \ldots = \sum_{k=0}^{\infty} x(kT)v^k \tag{2.13}$$

Readers with any experience in digital techniques will recognise eqn (2.13) as the definition of the z-transform – almost. In the argument above, v was used in the creation of the polynomial; obviously any letter of the alphabet would serve the purpose but it is conventional to use z and to raise it to *negative* powers, perhaps because delay is a more negative concept than advance.

Employing the accepted convention then, eqn (2.13) becomes eqn (2.14), the definition of the z-transform $x(z)$ of the sample sequence $x(kT)$:

$$x(z) = \sum_{k=0}^{\infty} x(kT)z^{-k} \tag{2.14}$$

In concept the summation in eqn (2.14) can be defined from $-\infty$ to ∞, rather than zero to ∞, so the definition given here is sometimes referred to as the 'unilateral z-transform'.

Although it has been stressed that one can only take the transform of a sequence of samples, not of a waveform, the temptation to speak of 'the z-transform of $x(t)$' is irresistible. The phrase will often be encountered and must always be interpreted as meaning 'the z-transform of the samples from $x(t)$'. In short, $\mathfrak{Z}\{x(kT)\}$ is correct but $\mathfrak{Z}\{x(t)\}$ and even $\mathfrak{Z}\{x(s)\}$ are used to mean the same thing.

2.3 The discrete transfer function

The z-transform is defined as an operation that can only be applied to a sequence of numerical values, yet the transformation of difference equations in mathematical texts is commonplace. How is this possible and what does it mean?

Consider equation (2.5), for example, rewritten as

$$\begin{aligned} & x_o(k) + b_1 x_o(k-1) + b_2 x_o(k-2) \\ & = a_o x_i(k) + a_1 x_i(k-1) + a_2 x_i(k-2) \end{aligned} \tag{2.15}$$

It involves three samples of two waveforms $x_o(t)$ and $x_i(t)$, but there are two ways of looking at it. The obvious one is that each side of the equation is the weighted sum of a number of samples from $x_i(t)$ or $x_o(t)$. But consider it another way. Take the left-hand side (say) as the weighted sum of samples from three identical waveforms x_o, x'_o, x''_o (see figure 2.7) each delayed by T sec with respect to its neighbour, but all sampled simultaneously. If $x_o(t)$ produces the sequence that transforms into $x_o(z)$, then $x'_o(t)$ leads to $z^{-1}x_o(z)$ and $x''_o(t)$ to $z^{-2}x_o(z)$ (compare eqns (2.11) and (2.12)).

The same argument applies of course to both sides of the equation leading to the transformation of eqn (2.15) to

$$x_o(z)(1 + b_1 z^{-1} + b_2 z^{-2}) = x_i(z)(a_o + a_1 z^{-1} + a_2 z^{-2})$$

and hence to the discrete transfer function:

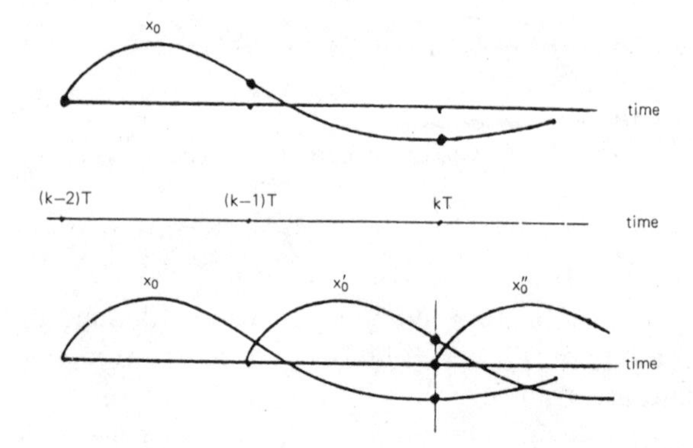

Figure 2.7 Samples from $x_o(t)$ and from delayed forms of $x_o(t)$

$$\frac{x_o(z)}{x_i(z)} = \frac{a_o + a_1 z^{-1} + a_2 z^{-2}}{1 + b_1 z^{-1} + b_2 z^{-2}} = F(z) \tag{2.16}$$

At this point consider the specific example of figure 2.4 again, in which the discrete transfer function is

$$\frac{x_o(z)}{x_i(z)} = \frac{2 + z^{-1} + z^{-2} - 4z^{-3}}{1 - z^{-1}}$$

$$= (2 + 3z^{-1} + 4z^{-2})$$

by long division.

It was found earlier (see eqn (2.6)) that the pulse response of the circuit is

$$x_p(kT) = \{2, 3, 4\}$$

and it is no coincidence that the result of the long division above is the transform of this. The reader is invited to compare carefully the steps in the division sum with the steps in the determination of the pulse response from figure 2.4 and it will be seen that they are the same; the division *must* give the transform of the pulse response.

Another way of looking at the situation is to note that a single pulse when sampled gives the sequence $\{1, 0, 0, \ldots\}$ and therefore has a z-transform of unity. When $x_i(t)$ is a pulse, the pulse response $x_o(t)$ must therefore have a transform $x_o(z)$ given by

$$x_o(z) = x_i(z)\, F(z)$$

$$= F(z)$$

since $x_i(z) = 1$.

Hence the expression for the transfer function must also be the expression for the pulse response.

2.4 The link between s and z

The operator z^{-1} implies a time delay of one sampling interval, T sec. A continuous function is similarly shifted when its s-transform is multiplied by e^{-sT} and thus z^{-1} in the z-domain maps into e^{-sT} in the s-domain, that is

$$z = e^{sT} \tag{2.17}$$

or

$$s = \frac{1}{T}[\ln z] \tag{2.18}$$

Any point in the s-plane may therefore be mapped into a point in the z-plane using eqn (2.17), and *vice versa* using eqn (2.18). Of particular importance are the points that lie on the imaginary axis of the s-plane: in this case $s = j\omega$ so that $z = e^{j\omega T}$, that is $|z|$ is unity. This means that the imaginary axis of the s-plane maps into a unit circle centred at the origin of the z-plane, where the stable region is the *interior* of the circle. This can be seen if a point $s = \sigma + j\omega$ is mapped across:

$$z = e^{(\sigma + j\omega)T} = e^{\sigma T}.e^{j\omega T} \tag{2.19}$$

The stable region of the s-plane is identified by a negative value for σ, in which case $|z| = e^{\sigma T}$ must be less than unity.

Finally it should be noted that the mapping is not unique. From eqn (2.19)

$$z = e^{\sigma T}(\cos \omega T + j \sin \omega T)$$

so a point P in the z-plane with cartesian co-ordinates ($e^{\sigma T} \cos \omega T$, $e^{\sigma T} \sin \omega t$) has the polar co-ordinates (r, θ) where

$$r = e^{\sigma T}$$

$$\theta = \omega T \tag{2.20}$$

Let P move anticlockwise along the arc of a circle from P′ to P″ in figure 2.8; θ changes from 0 to π rads while Q, the equivalent point in the s-plane with cartesian co-ordinates (σ, ω), moves from Q′ to Q″. When $\theta = \pi$ it can be seen from eqn (2.20) that $\omega = \pi/T$. Clearly therefore any point Q with an ω co-ordinate in the range $\pm\pi/T$ will map into a point P for which θ lies in the range $\pm\pi$. This identifies a portion of the s-plane known as 'the primary strip' and it is normally assumed when mapping from z to s that the s-plane point lies in it. It does not have to, of course. If θ for the point P″

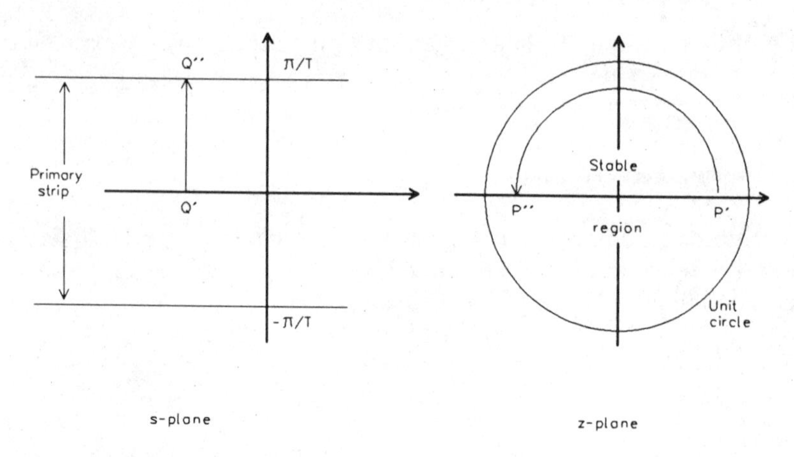

Figure 2.8 Defining the primary strip

were assumed to be 3π, 5π, etc. then ω for the point Q″ would be $3\pi/T$, $5\pi/T$, etc. In other words, there are an infinite number of points in the s-plane that map into any given point in the z-plane.

The many-to-one aspect of the mapping between the two planes is an important feature of the transition to a digital system and we will come across it again, in other forms. Its consequences prove to be important.

2.5 The z-transform table

In this section the transform table is discussed and explained, though the mathematics needed to derive it is not covered – the table is simply taken as given. The application of the table to the analysis of sampled data systems is the point.

The transform table (Appendix A) lists a number of functions of time $f(t)$ and the corresponding s- and z-domain functions $f(s)$ and $f(z)$ (including the modified z-transform whose use will be explained too). Note that $u_s(t)$ in the table represents the step function, which equals zero for $t < 0$ and unity for $t \geqslant 0$. Clearly $f(s)$ and $f(z)$ are transformations of the stated function of time, but let us be clear what this means, specifically in the case of $f(z)$, our central concern.

The function $f(t)$ is considered to be sampled at intervals of T sec to create $f(kT)$, an infinite sequence of sample values, from which is formed $f(z)$, an infinite series defined as before:

$$f(z) = \sum_{k=0}^{\infty} f(kT)z^{-k} \tag{2.21}$$

To form a table of infinite series *as such* would present obvious problems, but the problems are circumvented if the series are expressed in what is referred to as 'closed form' – the compact expressions given in the table.

When the tabled function $f(z)$ is expressed, still in closed form, in powers of z^{-1} instead of z, it takes a shape akin to $F(z)$ in eqn (2.16). If then we realise a network (figure 2.3) with a transfer function $F(z)$ equal to $f(z)$, that network will give an output $x_o(z)$ also equal to $f(z)$ when

$$x_i(kT) = \{1, 0, 0, \ldots\}$$

since

$$x_o(z) = F(z)x_i(z)$$

and

$$x_i(z) = 1$$

Example 2.1

To generate $f(t) = \sin 3t$ at intervals of 0.1 sec.

From the table it is seen that the corresponding function of z is

$$f(z) = \frac{z \sin(3T)}{z^2 - 2z\cos(3T) + 1}$$

Substituting $T = 0.1$ sec and writing in powers of z^{-1}:

$$f(z) = \frac{0.29552z^{-1}}{1 - 1.910673z^{-1} + z^{-2}}$$

$$= \frac{a_1 z^{-1}}{1 + b_1 z^{-1} + b_2 z^{-2}} = \frac{x_o(z)}{x_i(z)}$$

which leads to the discrete network of figure 2.9.

A single pulse at x_i generates at x_o the infinite sequence of values from the waveform, as the time-value table in figure 2.10 shows.

The derivation of closed forms for all the entries in Appendix A would require a knowledge of z-domain mathematics beyond the scope of this text, but there are many books available that cover it [Ho85, Og87, Fr90, Ph90 for example] and we therefore content ourselves with demonstrating how it may be done for three simple examples, to emphasise the nature of the table. Note that, in all cases, dividing out the closed-form expression gives the infinite series.

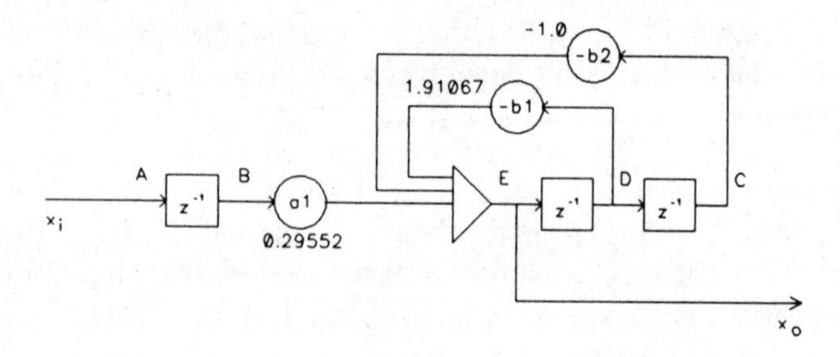

Figure 2.9 Generating the function sin 3*t* at intervals of 0.1 sec

sec	*A*	*B*	*C*	*D*	*E*
0	1	0	0	0	0
0.1	0	1	0	0	0.295520
0.2	0	0	0	0.295520	0.564642
0.3	0	0	0.295520	0.564642	0.783326

$E = 0.29552B + 1.91067D - C = x_o\,(kT) = \sin 3kT$

Figure 2.10 Time-value table for Example 2.1

(1) $f(t)$ is the unit step.
All samples have the value 1, therefore

$$f(z) = 1 + z^{-1} + z^{-2} + \ldots$$

To express this in closed form, note that

$$z^{-1}f(z) = z^{-1} + z^{-2} + z^{-3} + \cdots$$

so that

$$f(z) - z^{-1}f(z) = 1$$

Thus

$$f(z) = \frac{1}{1 - z^{-1}} = \frac{z}{z - 1}$$

(2) $f(t)$ is the unit ramp.
The sequence of sample values is

$$f(kT) = \{0, T, 2T, 3T, \ldots\}$$

so

$$f(z) = T(z^{-1} + 2z^{-2} + 3z^{-3} + \ldots)$$

$$z^{-1}f(z) = T(z^{-2} + 2z^{-3} + 3z^{-4} + \ldots)$$

Therefore

$$f(z)(1 - z^{-1}) = T(z^{-1} + z^{-2} + z^{-3} + \ldots)$$

$$= Tz^{-1}\left(\frac{1}{1 - z^{-1}}\right) \text{ from example (1)}$$

$$\therefore \quad f(z) = \frac{Tz^{-1}}{(1 - z^{-1})^2} = \frac{Tz}{(z - 1)^2}$$

(3) $f(t) = \mathrm{e}^{-at}$.
The sequence of sample values is

$$f(kT) = \{1, \mathrm{e}^{-aT}, \mathrm{e}^{-2aT}, \mathrm{e}^{-3aT}, \ldots\}$$

so

$$f(z) = 1 + \mathrm{e}^{-aT}z^{-1} + \mathrm{e}^{-2aT}z^{-2} + \ldots$$

therefore

$$\mathrm{e}^{-aT}z^{-1}f(z) = \mathrm{e}^{-aT}z^{-1} + \mathrm{e}^{-2aT}z^{-2} + \mathrm{e}^{-3aT}z^{-3} + \ldots$$

Hence

$$f(z)(1 - \mathrm{e}^{-aT}z^{-1}) = 1$$

$$f(z) = \frac{1}{1 - \mathrm{e}^{-aT}z^{-1}} = \frac{z}{z - \mathrm{e}^{-aT}}$$

Note that most of the tabulated functions cannot be determined so simply.

2.6 Use of the table

To explain the relevance of the z-transform table in the context of control, consider the simple sampled data system depicted in figure 2.11.

The plant is represented by $G(s)$ and is controlled by signals from the digital-to-analogue converter (DAC). These are generated by the digital processor, represented by $C(z)$, the input to which comes from the sampler or analogue-to-digital converter (ADC). It is often the case in practice that the A-to-D conversion takes place at the input $x_i(s)$ and in the feedback line so that the error x_e is digital rather than analogue; since we are not at this point discussing quantisation effects, such a scheme may be taken as identical conceptually to that shown in figure 2.11. The various elements of the loop are discussed again in greater detail below.

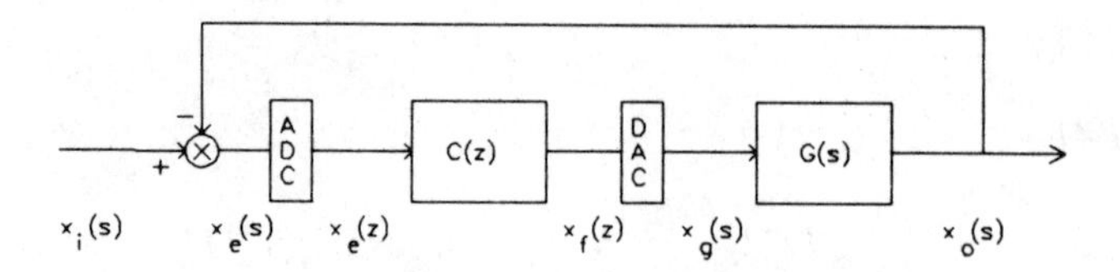

Figure 2.11 The fundamental sampled data system

When designing such a control system it is necessary to distinguish between two different but related tasks:

(1) to model the plant $G(s)$, *given that its input is provided by a digital processor*;
(2) to design the controlling software $C(z)$, given the plant model.

The first of these tasks is now considered in detail, the problem of designing the controller is deferred to the next chapter.

2.6.1 *The impulse*

To understand the sampled data loop and discuss the problem of devising a mathematical description of the plant within it, it will be found necessary to invoke the notion of an impulse, a concept that is often poorly understood. For example, two impulses, both spikes of infinite height and zero width, may have different strengths. How? What is the strength of an impulse anyway?

In order to clarify ideas, consider firstly a pulse P_1 of height H and duration T sec as in figure 2.12. The strength A of a pulse is defined as the area under it so in this case $A = TH$. If the height of the pulse is doubled and the duration halved to form the pulse P_2, and doubled and halved again to form P_3, it is obvious that the strength of all three pulses is the same; clearly if the doubling and halving is continued indefinitely, one forms an impulse with the same strength A. Figure 2.12 shows the

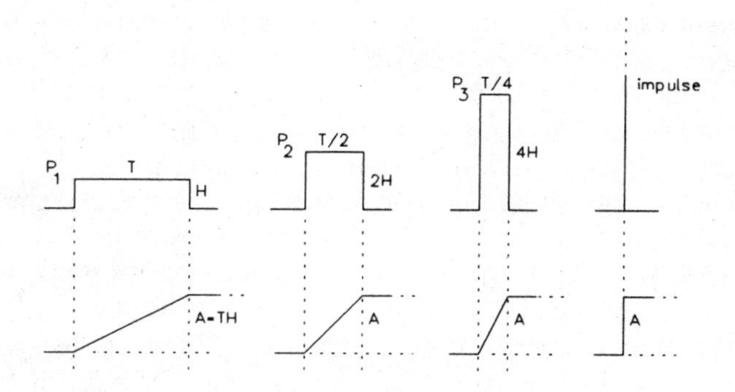

Figure 2.12 Transformation of a pulse to an impulse and its integral to a step

progression from pulse to impulse, and underneath, the integral of each pulse over time. The important things to note are that the final value of the integral is in all cases A, the strength of the pulse, and that in the limiting case the integral is a step function of height A.

The relevance of this to the plant model will shortly become clear.

2.6.2 *Mathematical relationships within the loop*

To discuss the operation of the system we need to know what form of 'data hold' is used in the DAC. Putting it another way, we need to know or to decide what is to happen between sampling instants to the analogue signal input to the plant. It is an arbitrary choice but almost always the chosen course of action is to maintain the voltage from the DAC constant at the latest value until replaced at the next sample instant, producing the familiar 'staircase' waveform. This form of reconstruction is carried out by a zero-order hold (ZOH). Occasionally one encounters a first-order hold, in which the DAC output ramps up or down over the sampling interval [Og87], but its use is rare and it will not be discussed here; higher orders still can be devised but they are not employed in practice.

What then is the sequence of events for which a mathematical description is sought? The error signal x_e (see figure 2.11), which is analogue and which we will presume to be a voltage, is sampled by the ADC to form the sample sequence $x_e(kT)$. This transforms into $x_e(z)$. Each sample of the sequence x_e captured by the processor is dealt with according to the algorithm defined by the discrete transfer function $C(z)$, to produce a sample of the sequence $x_f(kT)$; hence

$$x_f(z) = x_e(z)C(z) \tag{2.22}$$

In some instances we can realistically regard the time taken to sample x_e and compute x_f as negligible, and for the moment this will be assumed here. Thus a sample of x_e is taken and one of x_f produced, at the same sampling instant.

Let us suppose for the sake of the following argument that the current sample of x_f is the number A. What needs to be produced next is the link in mathematical terms between the number A and the voltage actuating the plant. The first step is a fictitious 'black box' that creates an impulse of strength A from the number; this is then followed by the ZOH which integrates the impulse to give a step of height A and then removes it after a delay of T sec to create a pulse of height A and duration T sec. The transfer function of the ZOH is hence given in the s-domain by the expression.

$$G_{zoh}(s) = \frac{1}{s}(1 - e^{-sT})$$

where e^{-sT} is the Laplace transformation of the time delay of T sec.

Thus successive values of x_f equal to A_1, A_2, etc. produce a succession of pulses of heights A_1, A_2, etc., forming x_g – the output from the DAC and input to the plant. Stated mathematically, x_g is the sum of these pulses, that is

$$x_g(s) = x_f(0)[G_{zoh}(s)] + x_f(T)[G_{zoh}(s)]e^{-sT} + x_f(2T)[G_{zoh}(s)]e^{-2sT} + \ldots$$

$$= \sum_{k=0}^{\infty} x_f(k)\left[\frac{1}{s}(1 - e^{-sT})\right] e^{-skT} \tag{2.23}$$

Now

$$x_o(s) = x_g(s)G(s)$$

which from eqn (2.23)

$$= \sum_{k=0}^{\infty} x_f(k)\left[\frac{1}{s}(1 - e^{-sT})\right] e^{-skT}\, G(s) \tag{2.24}$$

In eqns (2.22) and (2.24) we have therefore a description of the essential elements of the loop in mathematical terms, but the reader will note the uneasy conjunction of s- and z-domains. To be useful to us, both equations should lie in the same domain and hence the next step is to take the z-transform of the waveform $x_o(s)$. Note that the use of $x_o(z)$ does *not* imply the physical presence of a sampler acting on $x_o(t)$; one may choose to describe a continuous waveform in terms of z if it is sufficient to know it only at specific points in time.

Rearranging eqn (2.24):

$$x_o(s) = \left[\sum_0^{\infty} x_f(k)e^{-skT}\right][G(s)/s] - e^{-sT}\left[\sum_0^{\infty} x_f(k)e^{-skT}\right][G(s)/s] \tag{2.25}$$

Now

$$\mathscr{Z}\left\{\left[\sum_0^{\infty} x_f(k)e^{-skT}\right][G(s)/s]\right\} \triangleq x_f(z)\,\mathscr{Z}\{G(s)/s\}$$

since sampling a sum of waveforms is equivalent to summing the sampled waveforms. Hence the transform of eqn (2.25) is

$$x_o(z) = (1 - z^{-1})\,x_f(z)\,\mathscr{Z}\{G(s)/s\} \tag{2.26}$$

It is convenient to define

$$g_p(s) \triangleq G(s)/s$$

and

$$g_p(z) \triangleq \mathscr{Z}\{g_p(s)\}$$

so that

$$x_o(z) = (1 - z^{-1})x_f(z)g_p(z)$$

or

$$\frac{x_o(z)}{x_f(z)} = (1 - z^{-1})g_p(z) = G_z(z) \tag{2.27}$$

$G_z(z)$ is the discrete model of the plant and is described as its pulse transfer function.

Eqns (2.22) and (2.27) constitute the model we need of the elements of the sampled data loop.

Eqn (2.22) is straightforward but let us look again at eqn (2.27), the model devised for the plant. This is derived from the product of the original plant model $G(s)$ and the transfer function of the ZOH; when the plant with ZOH is subject to an impulse of strength $x_f(k)$ the result is a waveform the z-transform of which is given by eqn (2.27). To evaluate $G_z(z)$ in any particular case the steps are therefore to look up the z-transform of the step response of the given plant $G(s)$, that is, $g_p(z) = \mathscr{Z}\{G(s)/s\}$, and to multiply the resulting function by the factor $(1 - z^{-1})$, giving

$$\begin{aligned} G_z(z) &\triangleq (1 - z^{-1})\,g_p(z) \\ &= \left(\frac{z-1}{z}\right)g_p(z) \end{aligned} \tag{2.28}$$

This, the reader may remark, only gives the response of the plant to one impulse of unit strength. Remember however that the closed form expression found in the table of transforms defines a recursive algorithm, through its numerator and denominator coefficients (Example 2.1), which when stimulated by a single pulse generates the response of the plant to an impulse at x_f (to be formally correct, it generates the samples from the response). In reality this *is* what happens to the plant, except that it receives a new impulse at every sampling interval. If the model defined by $G_z(z)$, eqn (2.28), is supplied with the same sequence of pulses as the plant, it will therefore reproduce the plant output.

Example 2.2

To illustrate the application of the transform table, let us consider a plant described by

$$G(s) = \frac{3(s+2)}{s(s+4)}$$

It follows that

$$g_p(s) = \frac{3(s+2)}{s^2(s+4)}$$

$$= \frac{3/8}{s} + \frac{3/2}{s^2} - \frac{3/8}{s+4} \tag{2.29}$$

by partial fraction expansion.

From the table, substituting c for e^{-4T}:

$$g_p(z) = \tfrac{1}{8}\left[\frac{3z}{z-1} + \frac{12Tz}{(z-1)^2} - \frac{3z}{z-c}\right]$$

$$= \tfrac{1}{8}\left[\frac{3z(z-1)(z-c) + 12Tz(z-c) - 3z(z-1)^2}{(z-1)^2(z-c)}\right]$$

Letting $T = 0.1$ sec, say, so that $c = 0.67032$:

$$g_p(z) = \tfrac{1}{8}\left[\frac{z^2(2.18904) - z(1.79342)}{(z-1)^2(z-c)}\right]$$

$$G_z(z) = \left(\frac{z-1}{z}\right) g_p(z)$$

$$= \frac{z^{-1}(0.273630) - z^{-2}(0.224178)}{1 - z^{-1}(1.670320) + z^{-2}(0.670320)}$$

$$= \frac{a_1 z^{-1} + a_2 z^{-2}}{1 + b_1 z^{-1} + b_2 z^{-2}}$$

This expression defines the recursive network shown in figure 2.13 for which a time-value table is given in figure 2.14 where $F = 0.273630B - 0.224178C - 0.670320D + 1.670320E$ and the input (A) is a step, that is

$x_f(kT) = \{1, 1, 1, \ldots\}$

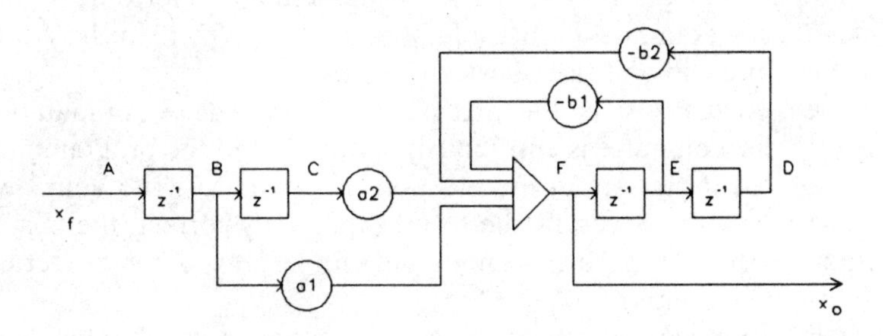

Figure 2.13 Discrete representation of the plant in Example 2.2

sec	A	B	C	D	E	F
0	1	0	0	0	0	0
0.1	1	1	0	0	0	0.273630
0.2	1	1	1	0	0.273630	0.506502
0.3	1	1	1	0.273630	0.506502	0.712053

Figure 2.14 Time-value table for Example 2.2

It can be seen that the output (F) is

$x_o(kT) = \{0, 0.273630, 0.506502, 0.712053, \ldots\}$

These figures can be checked because the response of the plant to a step is already known in s-domain form: it is $g_p(s)$, given above. In the time domain therefore, transforming eqn (2.29):

$$g_p(t) = 3/8 + 3t/2 - (3/8)(e^{-4t})$$

Substituting $t = 0, 0.1, 0.2$, etc., it will be found that the values quoted for x_o are correct.

2.7 Why use the transform table?

Values for $x_o(t)$ could be computed from the s-domain model using a suitable numerical algorithm to perform the integration (fourth-order Runge–Kutta, for example, to obtain high accuracy). Why then should we resort to the z-domain and the transform tables?

There are two reasons. In the first place the amount of computation performed by the computer is enormously reduced (and the programming greatly simplified if you are in the position of having to write your own integration algorithm). Secondly the result is precisely correct, there is no approximation involved, which cannot be said of any other numerical method, however complex.

We are able to use the transform table in this situation because the input to the plant is always a pulse; only the amplitude of the pulse changes from sample to sample. Thus the plant model performs a running summation of pulse responses of appropriate heights to determine the value of the output x_o from sample to sample.

2.8 Points of nomenclature

(1) Variables are denoted in lower case (for example, the error x_e, the step response of the plant g_p) and transfer functions in upper case (for example, the plant $G(s)$). The transfer function $G_z(z)$ is derived from $g_p(z)$ and is the z-transform of the response of the plant to a pulse at x_g (figure 2.11) or to a (fictitious) impulse at x_f; as such, it may also be regarded as a transfer function and written in upper case.

Given this convention, one must always ensure when referring to the z-transform $\mathfrak{Z}\{\cdot\}$ that the function represented by $\{\cdot\}$ is written in lower case; it is a function of time, or the s-transform of such a function, but *not* a transfer function.

(2) A waveform $f(t)$ which transforms into $f(s)$ is the impulse response of an analogue network with the transfer function $F(s) = f(s)$. The same waveform $f(t)$ may also be transformed into $f(z)$ and this is also the transfer function $F(z)$ of a discrete network which produces $f(kT)$ when stimulated by a single pulse. Some authors refer to a single pulse in the

discrete domain as 'an impulse' or 'a unit impulse' (though it is a somewhat unfortunate terminology) and for that reason a transformation listed in the transform table is sometimes called the Impulse Invariant Transformation (IIT).

(3) The subscript z, as in $G_z(z)$, will be used throughout this text with a special meaning, it indicates that the transfer function concerned embraces plant and data hold (normally a ZOH, though it does not have to be) and has been derived from the transform table, as was $G_z(z)$. For this reason a special term is used to refer to it – the pulse transfer function.

It is unfortunate that 'pulse transfer function' is used somewhat loosely in general, and in many cases means no more than 'discrete transfer function', but the definition given above is very useful, and since there is no alternative it is rigorously adhered to in this text.

2.9 Including time delay

Up to this point it has been assumed that negligible time elapses between sampling the error $x_e(t)$ and producing the corresponding input voltage to the plant $x_g(t)$. For example, in a ship autopilot sampling at 10-second intervals, the milliseconds spent on computation are obviously irrelevant. In an aircraft however, with the autopilot sampling, say, every 50 milliseconds, the computational time is a significant proportion of the sampling interval. It would therefore have to be included in the mathematical model of the system as 'dead time' or 'transport lag'.

A similar situation arises in many process plants, not because of computational delay but because a certain amount of transport lag is inherent in the process. In this case there is no delay in the determination of x_g, the delay occurs after that in the formation of x_o, but the effect on the model, on the computation of $x_o(z)$ from $x_e(z)$, is the same.

Note that although dead time has an adverse effect on the stability of any loop enclosing it, that is not our concern at the moment. We want at this point simply to adjust our mathematical model $G_z(z)$ so that the computed value of $x_o(kT)$ is correct in the new circumstances.

To do this the modelling process is begun in the manner already described, as far as the determination of $g_p(s)$. To find $g_p(z)$ from the transform table however, we refer to the column of *modified* z-transforms (Appendix A), that is

$$g_p(z) = \mathcal{Z}_m\{g_p(s)\}$$

The modified transform is seen to contain a parameter m; this relates to the amount of delay that must be accounted for. Let the dead time be T_d sec and let us state this as a proportion p of the sampling interval, that is

$p = T_d/T$

We define

$m = 1 - p$

and may then evaluate $G_z(z)$ from the table.

To understand the relevance of m, refer to figure 2.15. The curve y_c is identical to y_b except for a *delay* of $T_d = pT$ sec; similarly y_a is the same as y_b except for an *advance* of mT sec where $m = 1 - p$. Note that points P_1 and P_2 have the same ordinate values as R_1 and R_2.

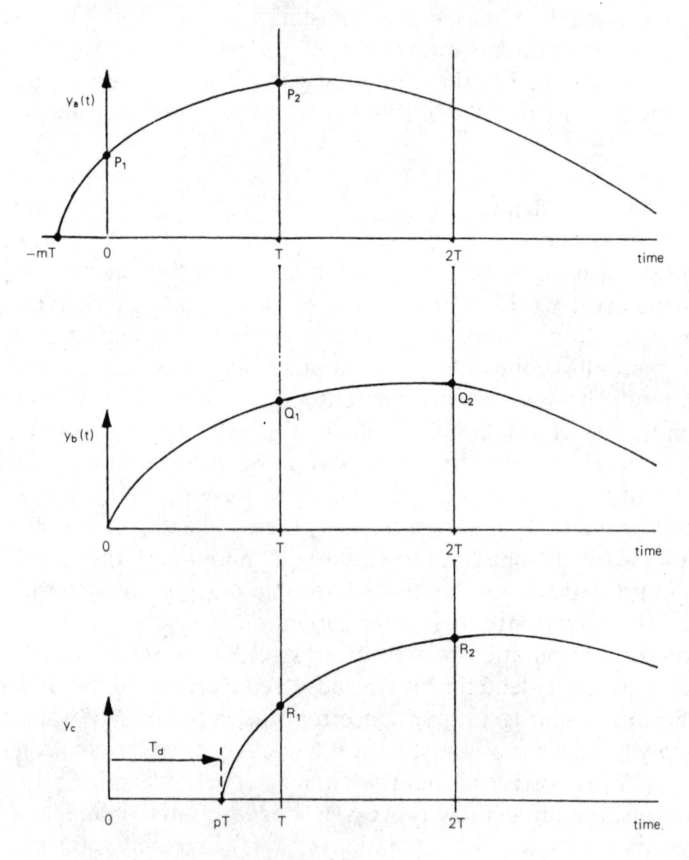

Figure 2.15 Explaining the modified *z*-transform

Use of the standard transform table gives sample values from the curve y_b (that is, Q_1 and Q_2) and one form of the modified table gives samples from y_a (that is, P_1 and P_2). However the modified table in Appendix A and in most texts on Control gives these values delayed by one sampling interval, that is, R_1 and R_2. Obviously the *z*-domain function generating P_1

and P_2 is related to that generating R_1 and R_2 merely by a factor z^{-1}, and both depend upon the value of m rather than p.

Example 2.3

To create a simple illustration of the use of the modified transform, the results of which can be easily verified, let us suppose that the plant is merely an integrator with some transport lag and that it is supplied with a step input from a digital processor. As in the last example therefore

$$x_i(kT) = \{1, 1, \ldots\}$$

The sampling interval is $T = 0.5$ sec and the dead time is $T_d = 0.2$ sec.

Under these conditions a model of the plant would be determined as follows:

$$G(s) = 1/s$$
$$\therefore\ g_p(s) = 1/s^2$$

Since $T_d = 0.2$ sec while $T = 0.5$ sec, therefore $p = 0.4$ and $m = 0.6$.

From the modified table:

$$g_p(z) = \frac{0.3}{z-1} + \frac{0.5}{(z-1)^2}$$

$$\therefore\ G_z(z) = \frac{z-1}{z}\, g_p(z) = \frac{0.3z^{-1} + 0.2z^{-2}}{1 - z^{-1}}$$

This is the discrete model of the plant.

For the constant input given above it is easily determined that the output sequence is

$$x_o(kT) = \{0, 0.3, 0.8, 1.3, \ldots\}$$

To verify, recall that the output is the integral of a unit step, that is, a ramp with unit slope, albeit delayed by 0.2 sec. See Figure 2.16. Clearly the first few samples of x_o taken at 0.5 sec intervals are as quoted above.

If the dead time is greater than the sampling interval, then it should be expressed in the form

$$T_d = (n + p)T$$

where n is an integer and p a fraction. The fractional portion of the delay is then treated in the manner described and the integer portion accounted for by multiplying the result by z^{-n}.

If for instance in this example the dead time were $T_d = 1.7$ sec, this would be rewritten $T_d = (3 + 0.4)T$ sec. Since p is again 0.4, the determi-

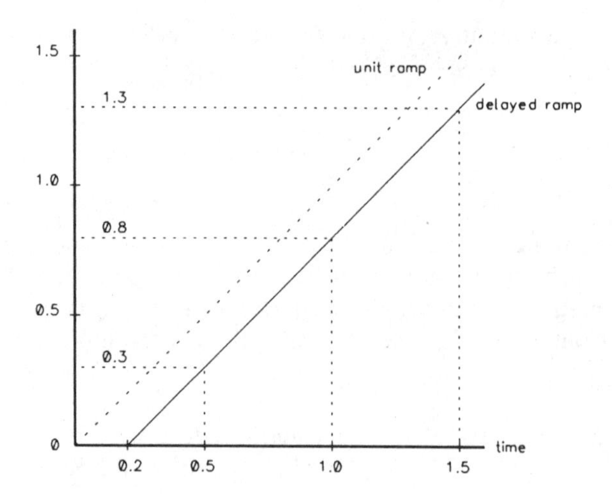

Figure 2.16 The output in Example 2.3

nation of $G_z(z)$ proceeds as before until the last line when we write

$$G_z(z) = z^{-3}\left[\frac{0.3z^{-1} + 0.2z^{-2}}{1 - z^{-1}}\right]$$

$$= \frac{0.3z^{-4} + 0.2z^{-5}}{1 - z^{-1}}$$

2.10 The impulse sampler and the star notation

It was found necessary above to invent, as part of the DAC, a non-existent 'black box' that produced an impulse of strength equal to the number being output from the processor. An alternative is to pretend that the sampler itself produces the impulses and that the processor is capable of dealing with them rather than with the numbers it handles in reality. This is in fact the conventional view: the sampler is an 'impulse sampler' (or 'ideal sampler') delivering an impulse equal in strength to the value of the sample from the waveform. The output from such an idealised device obviously bears no pictorial resemblance to any existing function of time, being a train of infinite spikes; that need not trouble us as long as the end result (the input to the plant) accords with our observations.

One may derive a mathematical description of the sampling process by regarding the sampled signal as the product of the analogue signal and a train of unit impulses. In the context of signal analysis and processing this might be a matter of some importance; here it is rather a digression, but some of the mathematics is nevertheless reviewed briefly in Appendix C. Included in the same appendix is an alternative view of the sampler which is sometimes preferred.

The waveform that is notionally output from the impulse sampler is a train of impulses of varying strengths and is by convention denoted with a star. Thus if the analogue signal before sampling is $x(t)$, the sampled signal derived from it and output from the sampler is $x^*(t)$. The Laplace transform of $x^*(t)$ is of course $x^*(s)$.

The train of weighted impulses is described mathematically

$$x^*(t) = x(0)\partial(t) + x(T)\partial(t - T) + x(2T)\partial(t - 2T) + \ldots$$

$$= \sum_{k=0}^{\infty} x(kT)\partial(t - kT) \tag{2.30}$$

where $\partial(t - kT)$ is the Dirac delta or unit impulse function. This is defined as zero except where the argument $(t - kT)$ equals zero, at which point it becomes a unit impulse. $x^*(t)$ is thus a series of impulses of strength determined by the input $x(t)$.

Since the Laplace transform of an impulse of strength A is the number A, we may transform eqn (2.30) thus:

$$x^*(s) = x(0) + x(T)e^{-sT} + \ldots + x(kT)e^{-skT} + \ldots$$

$$= \sum_{k=0}^{\infty} x(kT)e^{-skT} \tag{2.31}$$

If we make the substitution $z = e^{sT}$ then eqn (2.31) becomes

$$x(z) = \sum_{k=0}^{\infty} x(kT)z^{-k} \tag{2.32}$$

Eqn (2.32) is seen to be identical to eqn (2.14).

Since the star notation is applicable to transfer functions, $G(z)$ becoming $G^*(s)$ for instance, an equation in starred s-domain quantities is simply the transform of the equation in the z-domain. However computation in the s-domain with transcendental functions is not attractive, so calculations are invariably carried out in the z-domain and the star notation is retained purely for notational purposes. The presence of the star is used to declare unambiguously that the function is sampled; as pointed out before, use of the z-domain does not *per se* have this implication.

2.11 Analysis of sampled data systems

Since the natural description of a continuous or analogue element of a network is in terms of s while the digital elements are described in terms of

z there is an obvious problem of analysis: one may not mix functions of s and z in the same equation.

This situation has already been dealt with for the basic sampled data loop, but the question addressed here is wider. How are the principles explained for that simple case applied in the more general context of multiple loop systems with several s- and z-domain blocks?

The problem is examined by considering a series of examples of increasing complexity; the general principles emerge from consideration of the specific cases. Salient points are denoted by a †.

† *The first and most fundamental point to grasp is that if there is a digital element in the system then the input–output relationship for the system as a whole can only be expressed in terms of sampled quantities. It is important to recall that one may state the output of an analogue system as a function of z without implying that the output quantity is physically sampled.*

Example 2.4

The system shown in figure 2.17(a) is by now familiar territory but it is here the point of departure for unfamiliar regions, so we begin by considering the relevant mathematics again. The aim is simply to express the output in terms of the input.

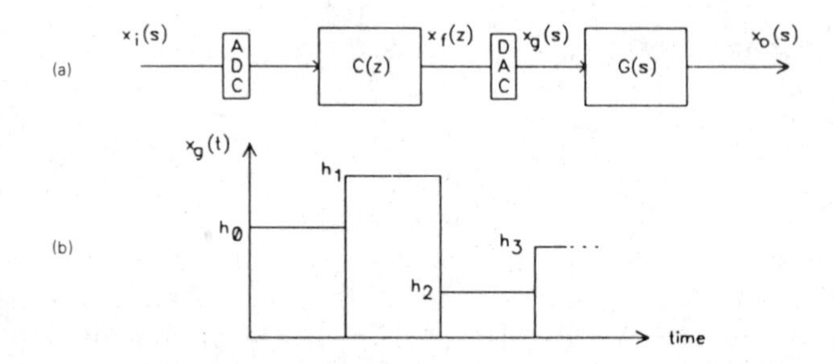

Figure 2.17 A sampled data system and the output from its DAC

The input to the plant, $x_g(s)$, is of necessity a series of pulses, a staircase that has been constructed using the ZOH. Successive pulses have heights $h_0, h_1, h_2, \ldots$, each delayed by T sec with respect to its predecessor (see figure 2.17(b)). Thus

$$x_g(s) = \frac{1}{s}(1 - e^{-sT})[h_0 + h_1e^{-sT} + h_2e^{-2sT} + \ldots] \tag{2.33}$$

The heights $h_0, h_1, h_2, \ldots$ are obtained from the processor and are given by the product

$$\begin{aligned} x_i(z)C(z) &= h_0 + h_1 z^{-1} + h_2 z^{-2} + \ldots \\ &= x_f(z) \end{aligned} \tag{2.34}$$

From $x_g(s)$, eqn (2.33), we derive $x_g(z)$. Note that the z-transform of a pulse is

$$\mathfrak{Z}\left\{\frac{1}{s}(1 - e^{-sT})\right\} = \frac{z}{z-1} - z^{-1} \cdot \frac{z}{z-1} = 1$$

and that

$$\mathfrak{Z}\{e^{-nsT}x(s)\} = z^{-n}\mathfrak{Z}\{x(s)\}$$

Hence, transforming eqn (2.33) one term at a time:

$$x_g(z) = h_0 + h_1 z^{-1} + h_2 z^{-2} + \ldots$$

This is seen in eqn (2.34) to be also the definition of $x_f(z)$. x_f, by its nature, is merely a set of numbers $\{h_0, h_1, h_2, \ldots)$ so its z-transform is the series given by eqn (2.34). x_g is a waveform, albeit a discontinuous one (the staircase), which sampled at intervals of T sec gives the same sequence of numbers as x_f. Hence $x_g(z)$ *should* equal $x_f(z)$ though $x_g(t)$ is not the same as $x_f(t)$. A little mathematics has been used to prove the obvious, but it may be useful to show that mathematics and common sense conform.

The output is $x_o(s) = x_g(s)G(s)$ so

$$x_o(z) = \mathfrak{Z}\{x_g(s)G(s)\}$$

where x_g is given by eqn (2.33). The z-transform of the product $x_g(s)G(s)$ may be written

$$\begin{aligned} x_o(z) &= \mathfrak{Z}\left\{\frac{G(s)}{s}[h_0 + h_1 e^{-sT} + \ldots]\right\} \\ &\quad - \mathfrak{Z}\left\{e^{-sT}\frac{G(s)}{s}[h_0 + h_1 e^{-sT} + \ldots]\right\} \\ &= \left[\mathfrak{Z}\left\{\frac{G(s)}{s}\right\}h_0 + \mathfrak{Z}\left\{\frac{G(s)}{s}\right\}h_1 z^{-1} + \ldots\right] \\ &\quad - \left[z^{-1}\mathfrak{Z}\left\{\frac{G(s)}{s}\right\}h_0 + z^{-1}\mathfrak{Z}\left\{\frac{G(s)}{s}\right\}h_1 z^{-1} + \ldots\right] \end{aligned}$$

$$= (1 - z^{-1}) \mathscr{Z} \left\{ \frac{G(s)}{s} \right\} [h_0 + h_1 z^{-1} + \ldots]$$

$$= \frac{z-1}{z} \mathscr{Z} \left\{ \frac{G(s)}{s} \right\} x_f(z) \tag{2.35}$$

by substitution from eqn (2.34). Finally eqn (2.35) is written

$$x_o(z) = x_f(z) G_z(z)$$

where

$$G_z(z) = \frac{z-1}{z} \mathscr{Z} \left\{ \frac{1}{s} G(s) \right\}.$$

Since $x_f(z) = x_i(z)C(z)$, the aim of relating input and output mathematically is accomplished:

$$x_o(z) = x_i(z)C(z)G_z(z)$$

or

$$\frac{x_o(z)}{x_i(z)} = C(z)G_z(z) \tag{2.36}$$

† Note that a plant $G(s)$ which derives its input from a ZOH must give rise to a discrete model of the form $G_z(z)$, incorporating the ZOH.

Example 2.5

Let us now increase the complexity of the network a little, to that of figure 2.18, and again express the output in terms of the input.

Since figure 2.18 between x_d and x_o is identical to figure 2.17(a), we may use eqn (2.36) to write

$$x_o(z) = C(z)G_z(z)\, x_d(z)$$

But

$$x_d(z) = \mathscr{Z}\{x_i(s)\, P(s)\} \tag{2.37}$$

which is denoted $XP(z)$, so

$$x_o(z) = C(z)G_z(z)XP(z) \tag{2.38}$$

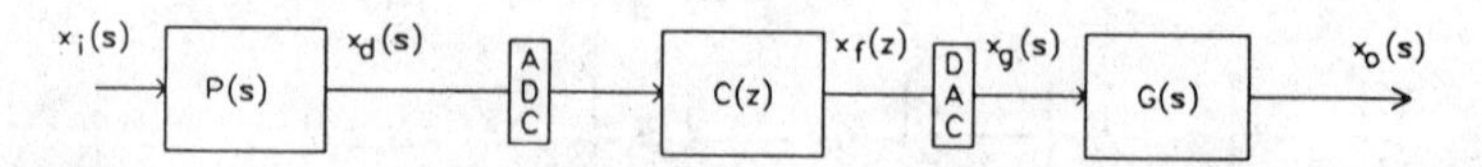

Figure 2.18 A slightly more complex system

Since $x_i(z)$ does not appear on the right-hand side of eqn (2.38), the input–output relationship in this example cannot take the form of a transfer function. Of course, we may still compute the output if we know the input; it is merely the compact form of expression provided by a transfer function that is missing. Instead we have to have *two* equations, (2.37) and (2.38) in this case.

† *When analogue elements are interposed between the input and the processor, the input–output relationship for a sampled data network cannot take the form of a transfer function.*

† *An expression such as* $\mathfrak{Z}\,\{x(s)G(s)\}$ *is denoted* XG(z). *It is not possible to extract* X(z) *from this expression, and* XG(z) *must be treated as an irreducible entity.*

Example 2.6

Figure 2.19 shows the system of figure 2.17(a) with the addition of feedback, and is of course the basic sampled data loop again.

† *From this point forward it is to be assumed that there is an ADC at the input to every z-block and a ZOH at its output. Concatenated z-blocks may simply be written as one block with the concatenated transfer function.*

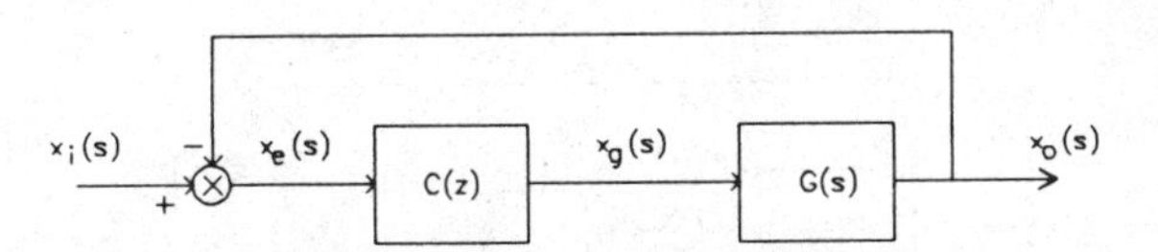

Figure 2.19 The basic sampled data loop again (remember that a *z*-block is assumed to have an ADC and a DAC at input and output)

With experience of previous examples, we may write down by inspection that

$$x_e(s) = x_i(s) - x_o(s)$$

$$\therefore \quad x_e(z) = x_i(z) - x_o(z)$$

$$x_g(z) = C(z)\, x_e(z)$$

$$x_o(z) = G_z(z) x_g(z)$$

Hence

$$x_o(z) = (x_i(z) - x_o(z))G_z(z)C(z)$$

$$\therefore \quad \frac{x_o(z)}{x_i(z)} = \frac{G_z(z)C(z)}{1 + G_z(z)C(z)}$$

Since the argument is z in all cases, we may drop it for convenience and write

$$\frac{x_o}{x_i} = \frac{G_zC}{1 + G_zC}$$

This example is quite straightforward and is inserted chiefly as a contrast with subsequent examples that may look almost as simple, but are not.

Example 2.7

The feedback network of figure 2.20 provides the next example. It shows a 'node' labelled x_a, which is a convention introduced at this point because it has not hitherto been necessary.

† *All signals that are input to a sampler are denoted by a node and labelled appropriately. A series of equations is then written down by inspection, the first (the output equation) being the link between the output and the nearest node upstream (that is,* against *the direction of signal flow). Succeeding equations (the nodal equations) relate only nodes to nodes and/or inputs.*

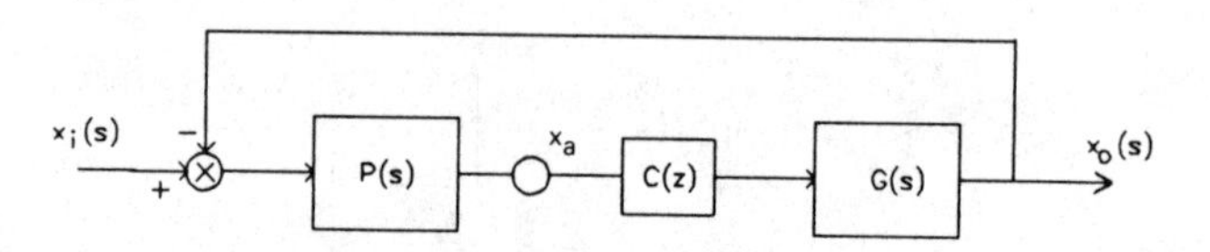

Figure 2.20 The system of figure 2.18 in a feedback loop

Note:

(a) the output will not normally be a node;
(b) the input may figure in the output equation along with one or more nodes, as the topology of the circuit requires (see Example 2.9);
(c) a nodal equation may relate a node to itself, eqn (2.40) below for instance;
(d) there are as many nodal equations as there are nodes, and output equations as there are outputs.

Accordingly we have in this example the following two equations:

$$x_o = x_a CG_z \tag{2.39}$$

$$x_a = (X_i P) - x_a C(GP)_z \tag{2.40}$$

where

$$(X_i P) = \mathfrak{Z}\,\{x_i(s)P(s)\}$$

$$(GP)_z = \frac{z-1}{z}\,\mathfrak{Z}\left\{\frac{1}{s}\,G(s)P(s)\right\}$$

The quantity $(GP)_z$ is defined as shown because that portion of x_a due to the feedback is derived from the pulse reponse of the composite block $G(s)P(s)$.

Eliminating x_a from eqns (2.39) and (2.40) leads to

$$x_o = \frac{(X_i P)CG_z}{1 + C(GP)_z} \tag{2.41}$$

No transfer function can be derived.

Example 2.8

The network of figure 2.21 contains an s-block in the feedback loop but not between input and processor, so a transfer function can be determined.

The necessary equations are

$$x_o = x_a CG_z$$

$$x_a = x_i - x_a C(GP)_z$$

and the transfer function is

$$\frac{x_o(z)}{x_i(z)} = \frac{CG_z}{1 + C(GP)_z}$$

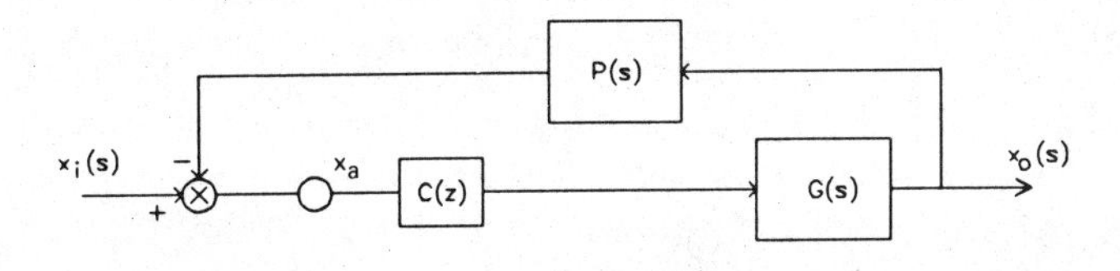

Figure 2.21 An example with an analogue network in the feedback loop

Example 2.9

The relevant network is that of figure 2.22 and the equations are

$$x_o = (X_iG) - x_aPG_z$$

$$x_a = (X_iGQ) - x_aP(GQ)_z$$

Thus

$$x_a = \frac{(X_iGQ)}{1 + P(GQ)_z}$$

and therefore

$$x_o = (X_iG) - \frac{(X_iGQ)PG_z}{1 + P(GQ)_z}$$

This result illustrates well how complex the computation of the output may be, even for a simple example. A great deal of arithmetic drudgery would clearly be involved in evaluating x_o given the functions P, Q, G and the input.

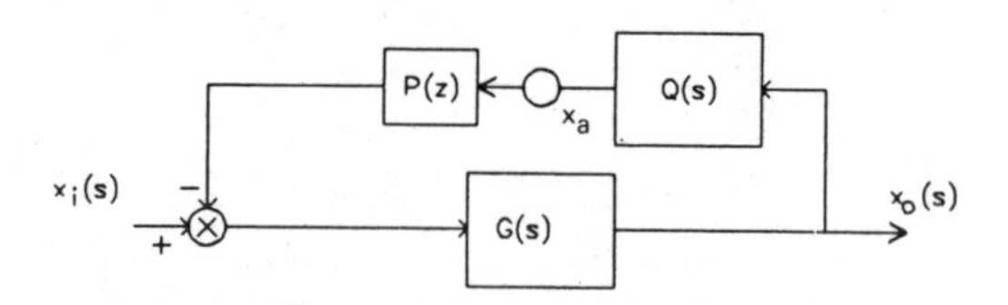

Figure 2.22 An example with both analogue and digital networks in the feedback loop

Example 2.10

Figure 2.23 shows a rather more complex network than hitherto. The expression for the output $x_o(z)$ would be enormous and there would be little profit in writing it out, but the equations from which it may be derived are

$$x_o = (X_iFG) - x_aD(FG)_z - x_bC(FG)_z$$

$$x_a = (X_iF) - x_aDF_z - x_bCF_z$$

$$x_b = (X_iFGB) - x_aD(FGB)_z - x_bC(FGB)_z$$

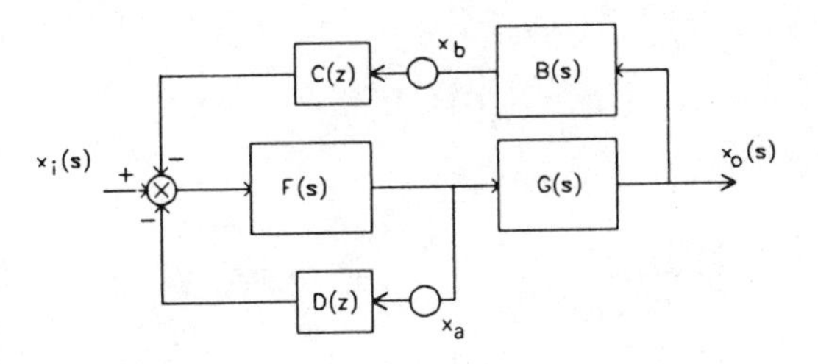

Figure 2.23 A rather complex sampled data system

Example 2.11

The network topology of figure 2.24 is simple enough but it gives rise to a degree of complexity. It may be dealt with in the same basic manner as the other examples but requires something extra. The input to the sampler is labelled x_a as normal, and the output equation fashioned to relate x_o to x_a and x_i. However there is a complete s-domain loop in the diagram, with transmittance $-F(s)G(s)$, so the contribution of $x_i(s)$ to $x_o(s)$ is given by

$$x_o'(s) = x_i(s)\frac{F(s)G(s)}{1 + F(s)G(s)} = x_i(s)\,J(s)$$

$$\therefore \quad x_o'(z) = X_iJ(z)$$

This circuit is regarded as broken at the point A and the signal x_a is treated as another input to the network, making a further contribution to the value of x_o. This too is complicated by the existence of the loop and is given by

$$x_o''(z) = x_a(z)C(z)\left[\frac{z-1}{z}\,\mathscr{L}\left\{\frac{1}{s}\,\frac{G(s)}{1 + G(s)F(s)}\right\}\right]$$

Letting

$$K(s) = \frac{G(s)}{1 + G(s)F(s)}$$

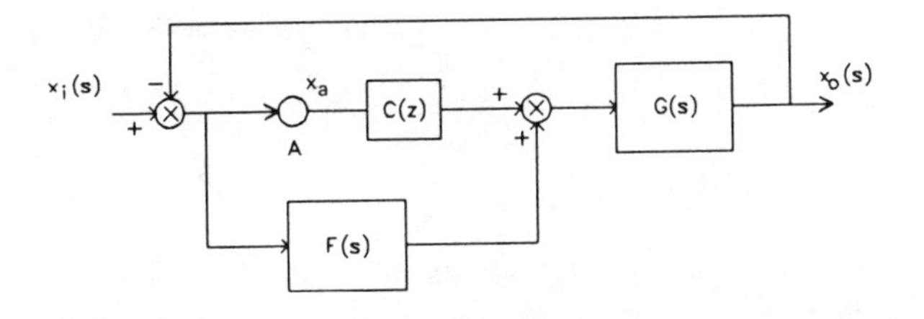

Figure 2.24 A system that looks simple but is quite complex analytically

the output equation may be written

$$x_o = x'_o + x''_o = (X_i J) + x_a C K_z \tag{2.42}$$

In a similar way $x_a(z)$ is defined by

$$x_a = \mathscr{Z}\left\{x_i(s)\frac{1}{1 + G(s)F(s)}\right\} - x_a C K_z$$

If

$$L(s) = \frac{1}{1 + G(s)F(s)}$$

$$x_a = (X_i L) - x_a C K_z \tag{2.43}$$

From eqns (2.42) and (2.43):

$$x_o = (X_i J) + \frac{(X_i L) C K_z}{1 + C K_z}$$

It would be possible to carry on with further examples, multiplying complexity upon complexity, but it is felt that the general principles have been adequately illustrated in the eight examples above. Further illustrations would probably show nothing new, though if a network is sufficiently complex the mechanics of applying the principles can present difficulty. For this reason a number of papers have been published over the years which treat the problem as one of signal flow graph, or connection matrix, reduction [Se67, Mu77, Sa81]. These methods have the advantage that they are quite general in application and will handle problems of any complexity; they have the disadvantage that understanding the method is a problem of the same order as understanding the analysis. Since the great majority of sampled data networks is of no great topological complexity, an understanding of the points made in this chapter should suffice for most purposes. Should it be necessary to resort to, say, the method proposed by Salehi, the material of this chapter will be found to provide the necessary background.

Problems

In the problems that follow, any discrete blocks shown in the diagrams are assumed to include an ADC at the input and a ZOH at the output.

1. Determine the value of the function

$$f(t) = 1 - e^{-t}$$

at intervals of 1 sec using the z-transform tables, and check the first few values.

2. Determine the value of the step response of the plant

$$F(s) = \frac{1}{1 + 0.25s}$$

at intervals of 0.1 sec using the z-transform tables. Use the functions of time in the table to validate the result.

3.
$$G(s) = \frac{10(s + 1)}{s + 10}$$

(a) Determine the step response of the plant $G(s)$ as a function of time.
(b) Use the z-transform tables to do the same thing in terms of z, for $T = 0.1$ sec.
(c) Check the values derived from (b) against (a), for the first few steps.

4.
$$F(s) = \frac{10}{s + 10} = \frac{x_o(s)}{x_i(s)}$$

If the input to the plant $F(s)$ is $x_i(t) = \sin \omega t$ with $\omega = 5$ rad/sec, determine the output $x_o(t)$ at $t = 0.9$ sec using
(a) the Laplace transform;
(b) the z-transform.

5.
$$F(s) = \frac{1}{s(1 + 1.5s)} = \frac{x_o(s)}{x_i(s)}$$

The input to the plant $F(s)$ is a pulse of height 1 unit and duration 2 sec.
(a) Use the z-transform tables to determine the value of the output at $t = 0$, 2, and 4 sec.
(b) Likewise discover the output value at $t = 0.5$ and 2.5 sec in two different ways.

6. (a) Using the Tustin transformation, obtain a discrete transfer function $G(z)$ from the continuous transfer function

$$G(s) = \frac{10}{s^2 + 7s + 10} = \frac{x_o(s)}{x_i(s)}$$

given that the sampling interval is 0.1 sec.

(b) Express the discrete transfer function as a time-domain relationship, and as a block diagram.

(c) From the block diagram, obtain the first three samples of the pulse response of $G(z)$ and hence using the z-operator obtain the first three samples of the response of $G(z)$ to the input sequence $\{4, 3, 2\}$.

7. (a) Obtain a discrete transfer function $G_z(z)$ to model the continuous transfer function

$$G(s) = \frac{10}{s^2 + 7s + 10}$$

given that $G(s)$ is the plant shown in figure Q7 and $T = 0.1$ sec.

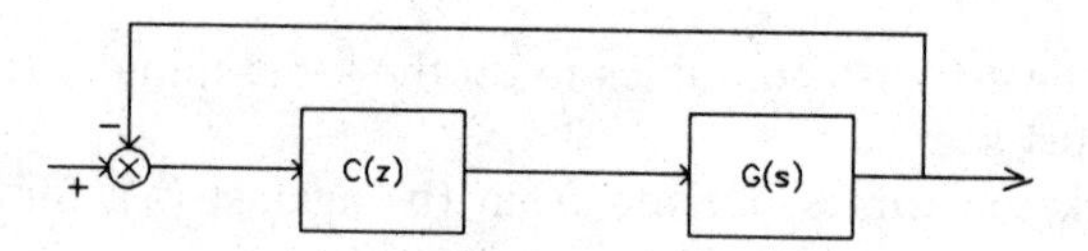

Figure Q7

(b) Explain why the discrete transfer function $G(z)$ obtained in **6**(a) above would be less satisfactory.

(c) Confirm your answer to (a) by determining the first few output samples of $G_z(z)$ when the input to the plant x_a is a step. Compare with the step response of $G(s)$.

8. A complex process plant may be described by the simple transfer function

$$F(s) = \frac{1}{1 + sT_p}$$

where $T_p = 1$ hour.

The input to the plant during a commissioning and testing exercise is a pulse $x_i(t)$ of height 1 unit and duration 2 hours. Show how the modified z-transform may be used to determine the value of the plant output at $t = 0.4$ hours, 1.4 hours, 2.4 hours, etc.; state the numerical values at 0.4 hours and 1.4 hours.

9. Figure Q9 shows a plant $H(s)$ with a digital processor $C(z)$ feeding back the derivative of the output $x_o(s)$:

$$C(z) = K(1 - z^{-1})/T$$

where K is an arbitrary gain and T is the sampling interval. The plant is described by

$$H(s) = \frac{1}{s + 1}$$

Determine the expression for $x_o(z)$ when the input $x_i(t)$ is a step function, and $K = 1$, $T = 1$ sec.

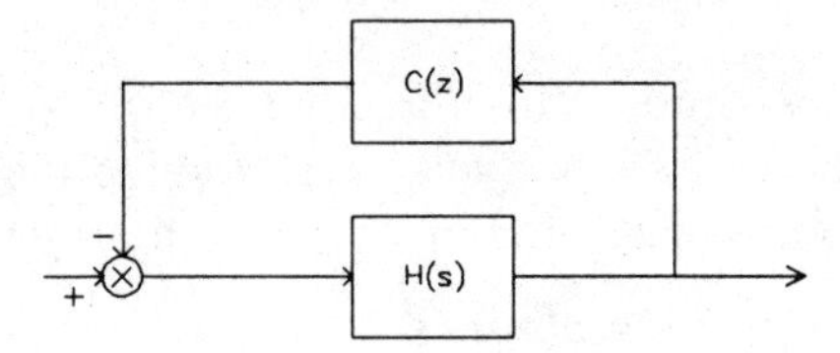

Figure Q9

10. The plant

$$G(s) = \frac{1}{1 + sT_p}$$

where $T_p = 4$ sec, is excited by the input $x_i(t) = e^{-t}$.
Determine the output of the plant at $t = 1$ sec and $t = 2$ sec using the z-transform tables.
Confirm your answer by reference to the relevant tabulated function of time.

11. Figure Q11 shows two sample data systems each comprised of a number of continuous and discrete blocks.
Determine the output $x_o(z)$ in each case.

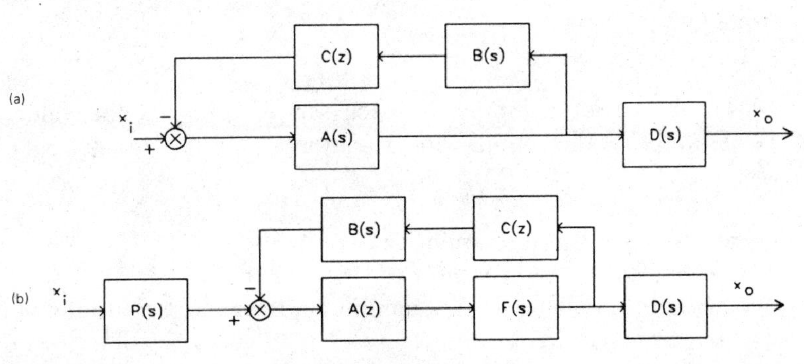

Figure Q11

12. In control studies to examine the performance of the braking system of a light-weight railway vehicle, the plant model used is that shown in figure Q12.

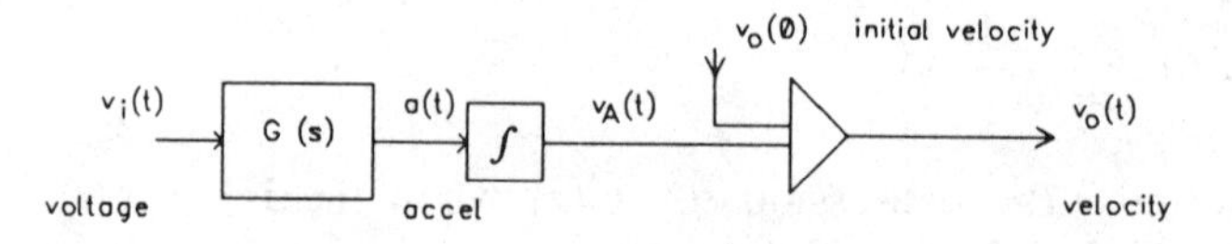

Figure Q12

The deceleration produced is directly proportional (constant of proportionality K) to the input voltage v_i, though the use of pneumatic actuators creates a dead time of T_d sec in the response. Hence

$$G(s) = -Ke^{-s}T_d = a(s)/v_i(s)$$

Knowing that the actuating voltage v_i is to be provided by a digital controller with a sampling interval T of 0.5 sec, draw up a suitable discrete model of the plant for use in the study when $K = 0.6$ m sec^{-2}/volt and the dead time T_d is (a) 0.2 sec, (b) 0.7 sec.
In both cases, demonstrate the validity of your model by considering the plant step response.

13. (a) In the circuit of figure Q13, express the relationship between the input x_i and the output x_o in terms of the operator z.

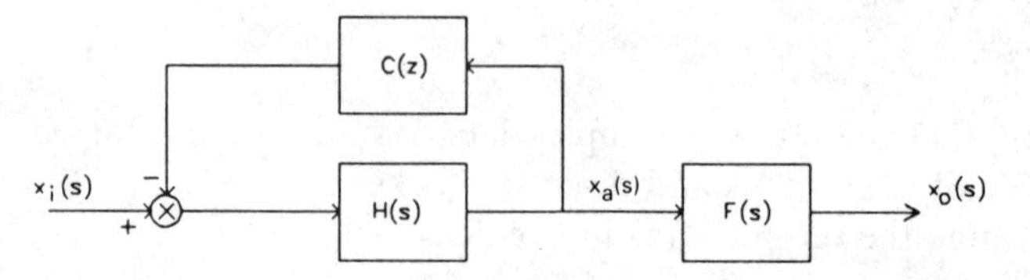

Figure Q13

(b) The processor computes at each sampling instant the value of dx_a/dt using the simple approximation

$$T\dot{x}_a = x_a(k) - x_a(k - 1)$$

where T sec is the sampling interval.
Given also that $H(s) = F(s) = 1/s$, show that when $T = 1$ sec and the input x_i is a step of unit amplitude, the expression found in (a) above reduces to

$$x_o(z) = \frac{0.5z^2}{(z - 1)^3}$$

(c) Show further that the output sequence in the step response begins

$$x_o(t) = \{0, 0.5, 1.5, \ldots\}$$

14. A plant $F(s)$ is controlled within a feedback loop by a digital processor that incorporates a zero-order hold in the DAC.

(a) Explain carefully why the Tustin transform, for example, may be employed in the emulation of the controller but the so-called impulse invariant transformation from the z-transform tables should be used to model the action of the plant.

(b) Given that $F(s)$ is described by

$$F(s) = \frac{4}{s(1 + 2s)}$$

use the transform tables to determine the z-domain model of the plant $F_z(z)$ for a sampling interval of 0.1 sec.

3 Controller Design

The chapter begins with some clarification of the concept of frequency response as applied to a digital algorithm. This entails an explanation of aliasing, and some words on the computation of the zero frequency gain. With an understanding of these basic ideas the reader is then equipped to consider the problem of designing a digital controller $C(z)$ for the sampled data loop (figure 2.11). The closed loop performance must meet some given specification and it is assumed that an adequate model of the plant $G(s)$ is available.

There are, broadly speaking, three approaches that one may take:

(1) Designing for continuous control but expressing the resultant s-domain function $C(s)$ in terms of z. The conversion from $C(s)$ to $C(z)$ is accomplished by any one of a number of simple techniques, all of which are approximations; several are discussed below. Emulation is the generic term for this procedure.

(2) Designing entirely in the z-domain. The plant model is converted from s to z in the manner described in Chapter 2 to give $G_z(z)$, so it incorporates the ZOH. The controller is then designed directly in the z-plane, using for example root locus techniques to assign z-plane poles and zeros to locations that will meet the closed loop performance specification.

(3) Designing in the w-plane. This is a hybrid scheme in which the plant is modelled as $G_z(z)$, as in method (2), but the model is then transferred to what is in effect the s-plane, where well-understood design methods can be employed. The transformation to the s-plane may be undertaken using any of the methods of emulation described in (1). However, since all such methods are approximations we have arrived, so the argument runs, in the 'nearly-s-plane'; the convention is to call it the w-plane and the transformation used to get there is normally the bilinear transform. Since the bilinear transformation is widely used, it is treated in some detail below.

Of these three methods, (1) is applied frequently and can be expected to give satisfactory results when the sampling rate is high (note that the ZOH is ignored); (3) has been found by experience to work very well across a wide range of sampling rates; (2) is not widely used because people do not in general like to undertake design on the unfamiliar z-plane.

Thus the first approach is worth considering because it is the simplest to undertake and can be very effective in the right circumstances, while the third is likely to work where the first one will not. There is therefore no particular need for method (2) which is more complex in computational terms than (1) and more difficult than (3) in application. For these reasons, emulation and w-plane design are described in detail but z-plane design is only touched upon. The latter does have its devotees, however, and the reader will find it covered well, in [Og87] for instance.

The chapter concludes with some brief consideration of the problems encountered in designing sampled data systems of more complex structures than the basic loop of figure 2.11. The section begins with a clear definition of what is involved in computing the frequency response of a system that comprises both analogue and digital blocks; after that it is necessary to consider how the topological complexity of the structure affects the process of designing the compensation.

3.1 The frequency response of a digital filter

Before going further it is necessary to say something about the frequency response of a digital filter, and it would be sensible to begin by clarifying the word 'filter'.

Control theory is now well established as a discipline in its own right, but it was developed in its early stages by engineers working in communications and signal processing. The term *filter* is simply borrowed from this activity and has degenerated to the point where it now means no more than transfer function or algorithm. There is not necessarily an implication that something unwanted is to be removed. Quite possibly, the widespread use of the term is due simply to its brevity.

3.1.1 Aliasing

Just as the frequency response of a function of s is obtained by setting $s = j\omega$ so the frequency response of a function of z is obtained by setting $z = e^{j\omega T}$. The resulting expression, for a given value of ω and T, evaluates to a complex number whose magnitude and phase can be plotted against ω and called the frequency response.

To this extent the frequency response of the digital filter is little different in concept from that of the analogue filter, but there is one important respect in which they are very different. The frequency response graphs tell us, in the case of a continuous filter, how a sinusoid will be altered in gain and phase on passing through the filter; the same graphs for the digital filter tell us how the samples from a sinusoidal waveform are affected. However the samples from a sinewave of any given frequency could also come from sinewaves of other frequencies. To see this clearly, suppose we sample a signal $x(t) = \sin \omega t$ at intervals of, say, 0.1 sec and obtain the sample sequence

$$x(kT) = \{0, 1, 0, -1, 0, 1, \ldots\} = \{\sin(\omega kT)\}$$

What is the frequency of the signal? The question may be answered by considering the sample for $k = 1$, that is

$$x(T) = 1 = \sin \omega T$$

or

$$\omega T = \sin^{-1}[1] \text{ rad}$$

The principal value of $\sin^{-1}[1]$ is of course $\pi/2$ rad, but there is in fact an infinite sequence of values:

$$\sin^{-1}[1] = \ldots, -\frac{7\pi}{2}, -\frac{3\pi}{2}, \frac{\pi}{2}, \frac{5\pi}{2}, \frac{9\pi}{2}, \ldots$$

$$= \frac{\pi}{2} \pm n2\pi \qquad n = 0, 1, 2, \ldots$$

Thus, from $\sin^{-1}[1] = \omega T$ rad

$$\therefore \quad \omega = \frac{1}{T} \sin^{-1}[1] \text{ rad/sec}$$

$$f = \frac{1}{2\pi T} \sin^{-1}[1] \text{ Hz}$$

$$= \ldots, -\frac{7}{4T}, -\frac{3}{4T}, \frac{1}{4T}, \frac{5}{4T}, \frac{9}{4T}, \ldots \text{ Hz}$$

$$= \ldots, -17.5, -7.5, 2.5, 12.5, 22.5, \ldots \text{ Hz}$$

with $T = 0.1$ sec.

In short, if the analogue waveform is

$$x(t) = \sin \omega t = \sin 2\pi ft$$

then the digital samples are

$$x(kT) = \{\sin 2\pi fkT\} \qquad k = 0, 1, 2, \ldots$$

$$= \left\{ \sin 2\pi\left(f \pm \frac{n}{T}\right) kT \right\} \quad n = 0, 1, 2, \ldots$$

The frequency of the sinewave from which the samples are taken is thus given by $f \pm nf_s$, where $f_s = 1/T$, an infinite number of possibilities spaced at intervals of f_s Hz. The physical significance of the negative frequency is that the sinusoid is inverted with respect to the positive frequency waveform, as seen in figure 3.1 ($\sin(-\omega t) = -\sin \omega t$).

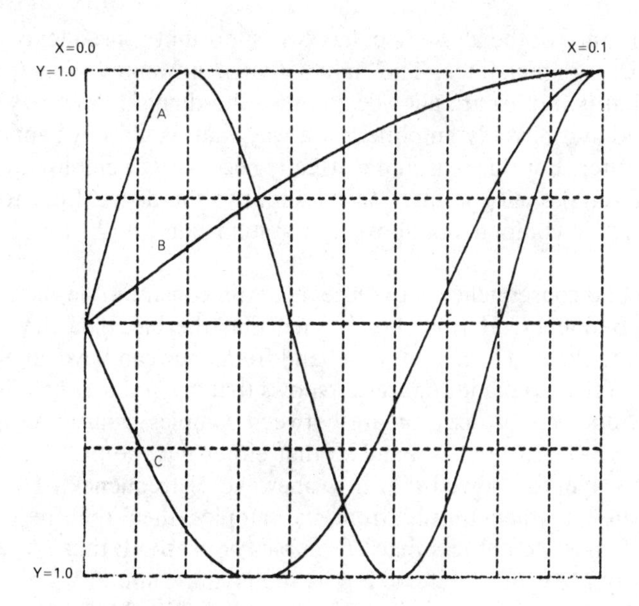

Figure 3.1 Three sinewaves yielding the same sample values at intervals of 0.1 sec. The frequencies of the curves are: (A) 12.5 Hz, (B) 2.5 Hz, (C) −7.5 Hz

The effect of producing multiple possibilities from the original sinewave is referred to as 'aliasing'. The component frequencies in a signal before digitisation are said to be in the 'baseband' and since each of these components is aliased on sampling, the baseband frequency response is repeated over the complete spectrum at intervals of $1/T$ or f_s Hz to give the frequency response of the sampled signal.

Bearing in mind that a sinewave of frequency $-f$ Hz is simply the sinewave of frequency f Hz inverted, it is clear that the attenuation at negative frequencies is the same as that at positive frequencies; by convention, therefore, the baseband is considered to embrace both positive and negative values of frequency, as shown in figure 3.2.

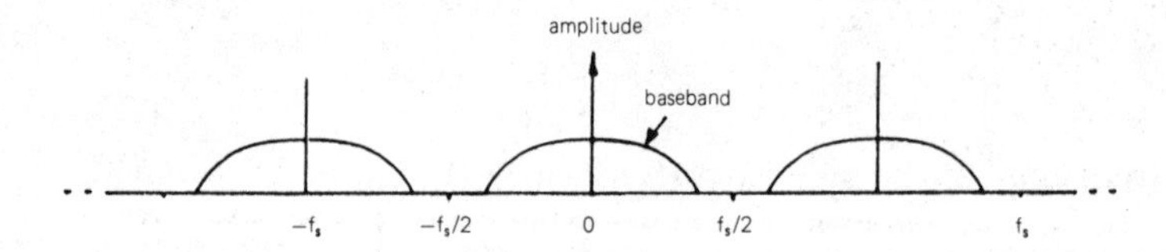

Figure 3.2 Frequency response of a hypothetical digital filter showing the effect of aliasing

The existence of the alias effect has two important consequences for the designer of a digital filter. The first is that if unwanted high frequency signals (that is, noise) are allowed to enter the digital filter, they may be transmitted and possibly amplified in a way that would not happen in an analogue filter. For this reason it is usually necessary to employ a low-pass (or antialiasing) analogue filter on the input to the digital filter to ensure that frequency components above the system bandwidth are effectively removed.

The second consequence is to place a certain constraint on the choice of sampling frequency. If f_s is chosen such that the baseband lies entirely within the range $\pm f_s/2$, then no baseband frequency can have an alias also lying within the baseband; figure 3.3 shows that when this constraint is *not* observed, one cannot discriminate between samples from two sinewaves both of which lie in the baseband so that either may be present.

In 1928, Nyquist showed that if a sinewave of frequency f Hz is to be reconstructed mathematically from its samples then the frequency of sampling f_s must be not less than $2f$ samples per second; thus $f_s/2 \geqslant f$, and the quantity $f_s/2$ is often referred to as the Nyquist limit.

In practice the value chosen for f_s is always considerably higher than that required by the Nyquist theorem, for reasons that will shortly be explained.

3.1.2 *Zero frequency response*

When the input to a filter is constant, the output, if the filter is not integrating, will also settle to a constant. The ratio of output to input in this circumstance, the zero frequency gain, may be of particular importance and is very easily determined. Suppose

$$\frac{y(z)}{x(z)} = F(z) = \frac{a_0 + a_1 z^{-1}}{1 + b_1 z^{-1}}$$

or

$$y(k) + b_1 y(k-1) = a_0 x(k) + a_1 x(k-1)$$

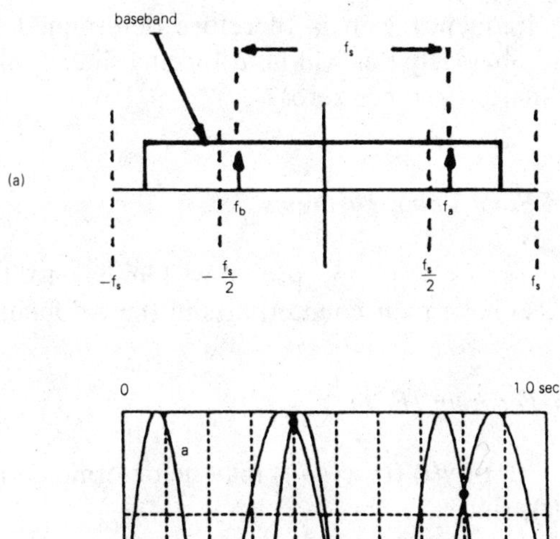

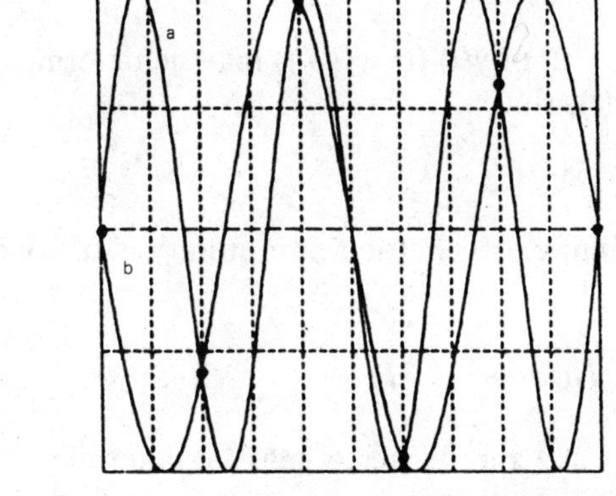

Figure 3.3 Illustrating the importance of the Nyquist limit. In diagram (a), f_s is 5 Hz so the Nyquist limit is 2.5 Hz; the baseband exceeds this and contains sinewaves of frequency $f_a = 3$ Hz and $f_b = -2$ Hz. These are graphed in diagram (b), at a and b respectively, and it can be seen that with $T = 1/5$ sec the samples from both sinewaves are identical. This happens when the separation between the two frequencies is f_s Hz

If the steady value of the input is denoted X and of the output Y, then all samples of the input and output in the steady state will have these values, that is

$$Y(1 + b_1) = X(a_0 + a_1)$$

or

$$\frac{Y}{X} = \frac{a_0 + a_1}{1 + b_1}$$

A simpler, though less physical, argument is simply to observe that $z = e^{j\omega T}$ when considering the frequency response, so that when $\omega = 0$,

$z = 1$. The zero frequency gain is therefore determined from $F(z)$ by setting $z = 1$, and obviously this will hold for any filter (unless by setting $z = 1$ the denominator becomes zero).

3.1.3 *Initial and Final Value Theorems*

Two valuable theorems that are probably familiar to the reader in s-domain terms also have their counterparts in the z-domain.

The Initial Value Theorem (IVT)

The initial value $y(0)$ of $y(t)$ (or $y(kT)$) may be determined from $y(z)$, the z-transform of $y(t)$, since

$$y(0) = \lim_{z \to \infty} \{y(z)\}$$

given that this limit exists. Setting z to infinity is of course setting z^{-1} to zero.

The Final Value Theorem (FVT)

The value of $y(t)$ as t approaches ∞ may be determined from $y(z)$, since

$$y(t)\Big|_{t \to \infty} = \lim_{z \to 1} \{(1 - z^{-1})y(z)\}$$

given that this limit exists.

These theorems often prove very useful and the FVT could have been employed to make the point in section 3.1.2 on the zero frequency response, as follows:

$$\frac{y(z)}{x(z)} = F(z) = \frac{a_1 + a_1 z^{-1}}{1 + b_1 z^{-1}}$$

If $x(t)$ is a step then

$$x(z) = \frac{z}{z - 1}$$

so that

$$y(z) = \frac{a_1 + a_1 z^{-1}}{1 + b_1 z^{-1}} \cdot \frac{z}{z - 1}$$

Applying the FVT to this definition of $y(z)$ gives the result quoted in section 3.1.2. The reader will find the IVT employed in Chapter 5.

3.2 Emulation

Emulation, to recapitulate, is the process of converting a transfer function in the s-domain to a transfer function in the z-domain (or *vice versa*). Be clear that this is an entirely different operation from the one described in Chapter 2 when the z-transform tables were used to create a discrete model of the plant. This is a point that is liable to cause confusion, so it is worth emphasising. The plant model in Chapter 2 was based upon the pulse response of the plant because the input from the ZOH is necessarily a sequence of pulses. In the more general case now being considered, the input to the s-domain transfer function block (no matter whether it represents a controller or anything else) is not fed from some digital element but is a continuous signal of arbitrary form. The modelling process used on the plant is clearly therefore not applicable to the controller.

So what *is* applicable? How might we transform $C(s)$ into $C(z)$ in some rational way, and what is the rationale? Several answers have been proposed and practised with varying degrees of success. What follows are a description of a number of the methods commonly encountered, an explanation of the thinking behind them, and brief comments on their effectiveness.

The effectiveness of any given method of emulation might be gauged in various ways but a common practice is to compare the frequency response of the s-domain and z-domain functions at some chosen sampling interval. If signal frequencies are low enough it will be found that all methods of emulation give results that compare well with the continuous function. As the frequency (of the signal) increases, disparity creeps in and it is possible to make distinctions between one method of emulation and another. In measuring or computing the frequency response of a digital filter, it makes no sense to proceed beyond the Nyquist limit, the frequency $\frac{1}{2}f_s$ Hz, and as one approaches this limit deviation from the ideal (continuous) filter response becomes great.

In practice, the function of z will cease to resemble the function of s well before the critical value of frequency. As a consequence of this, the sampling frequency must be chosen far above the bandwidth of the system to be controlled.

3.2.1 The bilinear transformation

At the beginning of Chapter 2 we looked at a simple scheme of digital integration, Tustin's method, that enabled us to change a differential equation into a difference equation. The reader may recall that the example used for illustration was the differential equation

$$\dot{x}_o + 2x_o = 3x_i \tag{3.1}$$

which may also be written in transfer function terms as

$$\frac{x_o(s)}{x_i(s)} = \frac{3}{s+2} = F(s) \tag{3.2}$$

Using Tustin's algorithm, eqn (3.1) became the difference equation

$$x_o(k) = a_0x_i(k) + a_1x_i(k-1) - b_1x_o(k-1) \tag{3.3}$$

where

$$a_0 = a_1 = 2T/(2+3T)$$

$$b_1 = -(2-3T)/(2+3T)$$

Hence

$$x_o(k) + b_1x_o(k-1) = a_0x_i(k) + a_1x_i(k-1)$$

and taking z-transforms

$$x_o(z)\ (1 + b_1z^{-1}) = x_i(z)\ (a_0 + a_1z^{-1})$$

or

$$\frac{x_o(z)}{x_i(z)} = \frac{a_0 + a_1z^{-1}}{1 + b_1z^{-1}} = F(z) \tag{3.4}$$

Thus, Tustin's algorithm has transformed $F(s)$ given by eqn (3.2) into $F(z)$ given by eqn (3.4).

The simplest way, arithmetically, to carry out this process is in fact to make the substitution

$$s = \frac{2}{T} \cdot \frac{1 - z^{-1}}{1 + z^{-1}} \tag{3.5}$$

in eqn (3.2), leading directly to $F(z)$ as defined in eqn (3.4). Eqn (3.5) is universally known as the bilinear transformation. Using the bilinear transform and employing Tustin's method or the trapezoidal rule are one and the same thing.

To appreciate how eqn (3.5) relates to the solution of a differential equation, let us look again at the trapezoidal rule.

Writing in general terms, suppose

$$y(t) = \int u(t)\ \mathrm{d}t \tag{3.6}$$

Using the trapezoidal rule, successive samples from the curve $u(t)$ define a trapezoid which represents the latest increment of area under the curve,

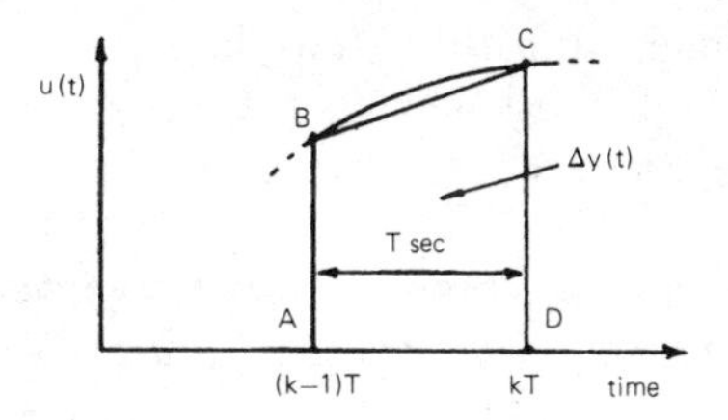

Figure 3.4 Illustrating trapezoidal integration

$\Delta y(t)$. See figure 3.4. Thus

$$\begin{aligned}\Delta y(t) &= y(k) - y(k-1)\\ &= \text{area ABCD}\\ &= \tfrac{1}{2}T[u(k) + u(k-1)]\end{aligned}$$

Taking z-transforms:

$$y(z)\,(1 - z^{-1}) = \tfrac{1}{2}T\,u(z)\,(1 + z^{-1})$$

giving

$$\frac{y(z)}{u(z)} = \frac{T}{2}\cdot\frac{1 + z^{-1}}{1 - z^{-1}} = G(z) \tag{3.7}$$

The differential equation (3.6) which may be written

$$\frac{y(s)}{u(s)} = \frac{1}{s} = G(s) \tag{3.8}$$

has been transformed into a difference equation which may be written as the discrete transfer function (3.7).

Equating the expressions (3.7) and (3.8) gives

$$\frac{1}{s} = \frac{T}{2}\frac{1 + z^{-1}}{1 - z^{-1}}$$

which is more often written

$$\begin{aligned}s &= \frac{2}{T}\cdot\frac{1 - z^{-1}}{1 + z^{-1}}\\ &= \frac{2}{T}\cdot\frac{z - 1}{z + 1}\end{aligned} \tag{3.9}$$

The name 'bilinear transform' merely alludes to the fact that both numerator and denominator are linear functions of z.

In applying the bilinear transformation therefore, the substitution given by eqn (3.5) is made on every occurrence of the Laplace operator; the resulting function is ordered in powers of z or z^{-1} to complete the transformation.

The bilinear transform is also applied in the reverse direction, to create a function of s from a function of z. In that case the substitution is

$$z = \frac{1 + sT/2}{1 - sT/2}$$

The experience of many people has been that emulation carried out through the bilinear transform is often very effective, though it is necessary that the sampling interval T should not be too large. The choice of sampling rate in a digital controller is a complex question that will be discussed in some detail in a later chapter, but as a rough guide it would be wise to aim for ten or twelve samples within the smallest system time constant if you intend to adopt a scheme of emulation. Another rule of thumb often used is to make f_s about ten times the system bandwidth.

It can be shown that mathematically one can reconstruct a waveform from its samples provided the sampling rate is such that at least two samples are available per cycle of the highest frequency component present in the signal. This is referred to as Shannon's Sampling Theorem, but it does not imply that the necessary mathematical processes can be implemented in a simple on-line algorithm. Shannon seems to have been the first person clearly to enunciate the principle in this context [Sh49] but in his paper he acknowledges the work of Nyquist [Ny28] by coining the term 'Nyquist limit' for the minimum sampling frequency.

3.2.2 Adams–Bashforth algorithms

A slightly more complex approach to integration is embodied in the Adams–Bashforth algorithms [Or81]. To obtain $y(t)$, the integral of $u(t)$, on this basis, one takes a number of samples of $u(t)$, say three for the sake of argument, and through them draws a smooth curve. It is then possible to compute the area under the curve between the limits $(k - 1)T$ and kT to arrive at a value for $\Delta y(t)$. The whole process boils down, on examination (see Appendix E, for example), to a simple algorithm which may be stated in the three-point case (described AB3) as

$$\Delta y(t) = (T/12)[5u(k) + 8u(k - 1) - u(k - 2)] \tag{3.10}$$

The notion may be applied to two samples of course but the result, as one might expect, is just the bilinear transform again; algorithms for four and more points exist but in this context are too cumbersome.

The algorithm AB3, as given by eqn (3.10), leads to

$$y(k) - y(k-1) = (T/12)[5u(k) + 8u(k-1) - u(k-2)]$$

and thus to

$$\frac{y(z)}{u(z)} = \frac{T}{12} \cdot \frac{5 + 8z^{-1} - z^{-2}}{1 - z^{-1}}$$

Hence the substitution

$$s = \frac{12}{T} \cdot \frac{1 - z^{-1}}{5 + 8z^{-1} - z^{-2}} \qquad (3.11)$$

can be tried in place of that given in eqn (3.9). Since (3.11) implies a quadratic curve rather than a straight-line boundary on $\Delta y(t)$ (the line BC in figure 3.4) it might be expected to represent some slight improvement on the bilinear transform. The evidence is that it rarely does, although it requires extra computational effort.

For example, if

$$G(s) = \frac{3(s+1)}{s+2}$$

the bilinear transform with $T = 0.05$ sec leads to

$$G(z) = \frac{2.92857 - 2.78571\, z^{-1}}{1 - 0.90476\, z^{-1}}$$

while AB3 (eqn 3.11) leads to

$$G(z) = \frac{2.940 - 2.7842\, z^{-1} - 0.012\, z^{-2}}{1 - 0.896\, z^{-1} - 0.008\, z^{-2}}$$

The extra numerator and denominator terms mean extra computation.

3.2.3 Use of the derivative

The Adams–Bashforth algorithms explained in the previous section attempt to improve on the bilinear transform by introducing a little extra complexity; in return for this it is hoped to gain a little more accuracy, though usually in vain.

Adopting the opposite philosophy, we might attempt to gain similar accuracy with reduced complexity – a worthwhile advantage – by using the substitution

$$s = \frac{1 - z^{-1}}{T} \qquad (3.12)$$

Eqn (3.12) derives from the simple observation that the first derivative, or the slope, of a curve at any point is given approximately by

$$\frac{\mathrm{d}x}{\mathrm{d}t} = \frac{1}{T}[x(k) - x(k-1)] \tag{3.13}$$

This approximation is a good one if T is small enough, so perhaps one might 'win' by using eqn (3.12) instead of (3.5) and sampling faster. Since there is less computation to be carried out during each sampling interval, this is feasible.

The verdict, however, seems to be that in most circumstances the bilinear transform is to be preferred.

3.2.4 *Pole and zero mapping*

The methods described up to this point have all had some basis in numerical mathematics. The idea behind pole and zero mapping is rather different. Since every point in the s-plane has a counterpart in the z-plane, through the mapping $z = \mathrm{e}^{sT}$, why not form $F(z)$ from $F(s)$ by mapping the poles and zeros across from one plane to the other?

Thus if

$$F(s) = \frac{K_s(s-a)}{(s-b)}$$

then

$$F(z) = \frac{K_z(z-c)}{(z-d)}$$

where $c = \mathrm{e}^{aT}$, $d = \mathrm{e}^{bT}$.

The value of K_z is chosen to ensure that the gain of $F(z)$ is identical to that of $F(s)$ at some specific point, very often zero frequency. In that case, we may say in this example

$$\frac{K_s a}{b} = \frac{K_z(1-c)}{(1-d)}$$

that is

$$K_z = \frac{K_s a(1-d)}{b(1-c)}$$

The notion of pole and zero mapping seems eminently sensible and appeals to many, but it raises the question of what to do about 'missing zeros'. $F(s)$ will often have fewer zeros than poles, with the zeros that are apparently missing located in fact at an infinite distance off in the left half plane. Where are these zeros to be placed in the z-plane? Some arbitrarily advocate the point $z = 0$, some $z = -1$; neither seems entirely satisfactory, and in truth the method does not work as well as might be expected even without a 'zero problem'.

3.2.5 *Bilinear transform with pre-warping*

The frequency response of a transfer function is obtained by substituting $j\omega$ for s in $F(s)$, or $e^{j\omega T}$ for z in $F(z)$.

When the bilinear transform is used to form $F(z)$, the proper transformation $s = (1/T)\ln z$ is replaced by the approximation

$$s = \frac{2}{T}\cdot\frac{z-1}{z+1}$$

(This was explained earlier by reference to Tustin's algorithm, but it is interesting to note the link with one of the power series expansions for the logarithm, namely

$$\ln x = 2[X + X^3/3 + X^5/5 + \ldots]$$

where $X = (x - 1)/(x + 1)$.
Taking only the first term in the series leads to the bilinear transform.)

Writing $e^{j\omega T}$ for z in the expression

$$\frac{2}{T}\cdot\frac{z-1}{z+1}$$

leads to

$$\frac{2}{T}\cdot\frac{e^{j\omega T}-1}{e^{j\omega T}+1}$$

$$= \frac{2}{T}\cdot\frac{e^{j\omega T/2}-e^{-j\omega T/2}}{e^{j\omega T/2}+e^{-j\omega T/2}}$$

$$= j\frac{2}{T}\tan\left(\frac{\omega T}{2}\right)$$

Thus to obtain the *discrete* frequency response from the *continuous* transfer function, we substitute

$$s = j\frac{2}{T}\tan\left(\frac{\omega T}{2}\right) \tag{3.14}$$

in $F(s)$.

Obviously when ω is small, $\tan(\omega T/2) \simeq \omega T/2$, and the difference in discrete and continuous response will be negligible.

As the frequency increases, however, the difference becomes rapidly more marked: the response of the continuous filter at $\omega \to \infty$ is seen from eqn (3.14) to be reproduced in the digital filter as $\omega = \pi/T$, a gross distortion. The periodic nature of the tan function in eqn (3.14) is yet another reflection of the periodic nature of the discrete frequency response.

The idea of pre-warping depends on the observation that it is possible to force the digital filter to have the same gain as the analogue filter at a chosen frequency by warping, or changing, the original function of s in the appropriate way.

The process will become clear with an example. Let

$$F(s) = \frac{10}{s + 10}$$

and suppose that we want the digitised filter to have the same gain and phase at $\omega = \omega_c$ as does the analogue filter (ω_c is referred to as the critical frequency of the warp).

To compute the frequency response of the analogue filter at $\omega = \omega_c$, s is replaced by $j\omega_c$. To compute the frequency response of the digital filter at $\omega = \omega_c$, when the bilinear transform has been used to form it, s is replaced by $j(2/T)\tan(\omega_c T/2)$ as shown above. If the responses of the two filters are to be identical at this frequency then s in the original filter must now be multiplied by a constant K such that

$$K\left[j\frac{2}{T}\tan\left(\frac{\omega_c T}{2}\right)\right] = j\omega_c$$

that is

$$K = \frac{\omega_c T/2}{\tan\left(\dfrac{\omega_c T}{2}\right)} \tag{3.15}$$

The bilinear transform is then applied, not to $F(s)$ as originally given, but to

$$F'(s) = \frac{10}{Ks + 10}$$

where K is defined in eqn (3.15).

It is easily appreciated that the same reasoning applies to any transfer function, however complex; s, on every occurrence, is to be multiplied by the factor K, s^2 by K^2 and so on. It is an important point, that this process leaves the zero frequency gain unaltered.

To complete the current example let $\omega_c = 10$ rad/sec (the corner frequency of the analogue filter) and $T = 0.1$ sec. Hence

$$K = \frac{0.5}{\tan(0.5)} = 0.915$$

The filter to which the bilinear transform is applied is therefore

$$F'(s) = \frac{10}{0.915s + 10} = \frac{10.929}{s + 10.929}$$

3.2.6 *A pole mapping technique*

In section 3.2.4 the technique of *pole and zero* mapping was described, and considered to be something of a disappointment. The method described now employs pole but not zero mapping; to form the zeros of $F(z)$ a more complex procedure is undertaken whereby the past samples of the input, $y(t - T)$, $y(t - 2T)$, etc. are expanded in a Taylor series. When the terms of this series are summed and equated with equivalent terms in the given differential equation, a set of algebraic equations is formed. The solution of these equations results in the numerator coefficients of $F(z)$. Appendix E explains the method more fully.

While the reasoning behind this technique is relatively complicated, the arithmetic complexity in its application is no greater than that of other methods. Since the method is of recent origin, experience with it is limited, but it appears to offer a more faithful transformation between the s- and z-domains than any of the alternatives.

There is one feature of this method which distinguishes it from all the others. It enables the user to compute from the set of input samples an estimate of the output quantity *at any moment of time*, extrapolating into the future or interpolating between past samples. One can, for example, fashion an algorithm to allow for the computational delay.

3.2.7 *Comparisons*

Other methods of emulation besides the ones above have been devised and employed, but those described are probably the most noteworthy.

The bilinear transform is without doubt the most commonly used; as well as being simple in conception it is effective in most situations. It is difficult to be more definitive in discussing methods of emulation because although one can undertake a formal and detailed comparison of the performance of a number of methods in a given situation, the result is only valid for the conditions obtaining in the experiment. One interesting example of such a trial is reported in [Ka81], for instance, where it is concluded that the most effective digital controller, from a group of eight, is provided by the bilinear transform.

There *are* situations in which other choices would be preferable. If for example it was necessary to design a notch filter in discrete form, one could emulate the continuous filter using the bilinear transform with pre-warping. That would be an appropriate choice because one particular frequency is of paramount importance.

In fact the technique of pre-warping is so simple that it is probably a good idea to check whether it holds any advantage in your situation or not. To recall, after choosing the sampling period T and the critical frequency ω_c at which the discrete and continuous filters should match, one sets

$$K = \frac{\omega_c T/2}{\tan(\omega_c T/2)}$$

and replaces s by Ks throughout the transfer function. The advantages, while uncertain, are worth investigating at little trouble.

In digital simulation of an entirely continuous dynamical system, one can solve the differential equations using complex numerical methods, but it is also common practice to use the bilinear transform, taking smaller steps in the time frame. However use of the pole mapping technique described in section 3.2.6 would probably give better results than the bilinear transformation. It is also a better way of emulating the notch filter mentioned above [Fo85a].

Finally, figure 3.5 offers a brief comparison in frequency responses between the analogue filter

$$F(s) = \frac{10}{s^2 + s + 10}$$

and three digital equivalents derived via the bilinear transform, the bilinear transform with pre-warping, and the pole mapping method used with the Taylor expansion. The centre frequency for the warping was $\omega_c = 0.5$ Hz or π rad/sec (almost exactly the frequency of peak gain in $F(s)$) and the sampling interval in all cases was $T = 0.4$ sec; this corresponds to about 5 samples per cycle at the natural frequency of the filter – a low value.

Only the gain/frequency curves are shown but they establish clearly an order of preference if that is based on fidelity to the analogue frequency response. The phase/frequency curves lead to the same conclusion, namely

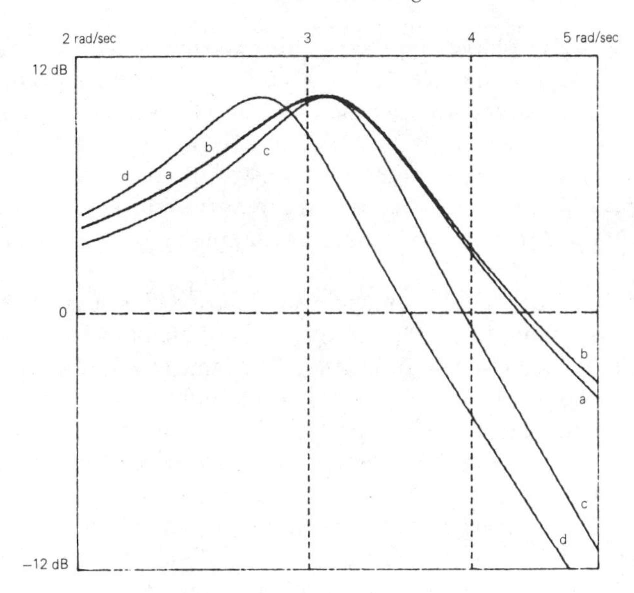

Figure 3.5 A comparison in emulation methods. The curves are identified as follows: (a) continuous response, (b) pole mapping with the Taylor expansion, (c) bilinear transform with pre-warping, (d) bilinear transform. $T = 0.4$ sec in all cases

that the best method is pole mapping with the Taylor expansion, and that the least accurate of the three is the plain bilinear transform.

It is worth noting, however, that one can argue for a more subtle basis of comparison and obtain different results (see for example [Ka81, Fo85b]).

3.3 Controller design on the *w*-plane

When the plant under control is fast-acting, the plant time constants are short and it may not be feasible to sample quickly enough to justify design through emulation. The best of the methods available in this circumstance, as explained at the beginning of the chapter, uses the so-called *w*-plane. Our purpose now is to consider what this means and how it is put to use.

The key deficiency in emulation methods is that they fail to recognise the presence of the ZOH. The smooth output of the analogue controller is replaced in the digital case, when sampling fast, by a curve of tiny steps that can be viewed as a smooth curve with small-amplitude, high-frequency noise superimposed. The plant cannot respond to such frequencies and therefore forms a natural filter, removing the noise and responding only to the underlying waveform.

With coarse steps of large amplitude, this reasoning obviously does not apply; equally clearly, the first thing to be done in coping with long sampling intervals is to incorporate the ZOH into the design process. This is achieved by forming a plant model in the manner of Chapter 2, designated $F_z(z)$ where the continuous model of the plant is $F(s)$. Such a model is appropriate because the plant receives from its controller only a series of flat-topped pulses of duration T, assuming that the DAC employs a ZOH. The input to the plant is always a pulse.

If you are experienced in s-plane root locus techniques it may be worthwhile at this point to attempt design on the same basis in the z-plane, method (2) at the beginning of the chapter, but some trial-and-error will be needed to achieve a similar degree of skill. Otherwise the next step is to transfer the z-plane model to the s-plane, simply to avoid the necessity of further work in the z-domain. It may well be no more than a matter of personal preference to avoid z-plane design, but it is the preference of many people. Transferring $F_z(z)$ to the s-plane is emulation again, but in reverse. Precisely the same problems as those discussed earlier present themselves; exact transcription from one plane to the other is not an option and the commonest substitute is again the bilinear transform. The technique of pole mapping with Taylor expansion may do the job better, but very good results can be achieved with the bilinear transform; any emulation method may of course be used.

Since emulation necessarily produces an approximation it has become the custom to refer to the result as a function of w and to speak of the w-plane rather s and the s-plane. This is mathematically more correct since the result of the transformation is not quite s, but why we should choose on this occasion to be quite so fastidious is something of a mystery. People have been using the bilinear transform for decades to travel from s to z without feeling it necessary to acknowledge the inaccuracy involved by inventing another letter for z. However, $F_z(w)$ is now treated exactly as if it were a function of s and the design from this point is undertaken in the classical s-domain manner. Any or all of the well-established frequency domain or root locus design methods may be employed to arrive at a suitable compensator $C(w)$ which the user feels will meet the specification. The final step is to transfer back to the z-plane, using the chosen method of emulation, to form $C(z)$ and hence the expression giving the current output of the processor.

On a point of nomenclature, the w-plane is referred to as the w'-plane in some texts ([Le85] for instance). The reason for the confusion is that in communications theory the original definitions were

$$w = \frac{z - 1}{z + 1}$$

and

$$w' = \frac{2}{T} \cdot \frac{z-1}{z+1}$$

However w defined in that way is rarely used in control. Increasingly, therefore, control textbooks have taken to defining

$$w = \frac{2}{T} \cdot \frac{z-1}{z+1}$$

with the dash (′) omitted.

Example 3.1

Let the plant to be controlled have the transfer function

$$F(s) = \frac{1}{s(1+10s)}$$

The controller placed in cascade is a simple phase advance network of the s-domain form

$$C(s) = G\left[\frac{1+skT_c}{1+sT_c}\right] \tag{3.16}$$

and the constants G, k and T_c should be chosen to give the compensated system a phase margin of 45°, with a crossover frequency of about 1 rad/sec. The problem is firstly tackled in analogue terms, then in digital form using emulation, and finally the design is undertaken on the w-plane. Any classical design technique would suit, but the one chosen for this illustration is frequency response shaping on the gain-phase plane (the Nichols chart). Those unfamiliar with the use of the gain-phase plane are referred to Appendix D, but some commentary is added here.

The analogue design. A feedback loop closed around the uncompensated plant gives a very poor closed loop response, since the phase margin is inadequate. Figure 3.6(a), (b) shows the step response of the closed loop and the frequency response locus on the Nichols chart for the open loop, before the addition of compensation. To increase the phase margin to 45° will impart much greater damping, and if the crossover frequency is raised from its original value of 0.3 to 1.0 rad/sec, the speed of response will be correspondingly increased.

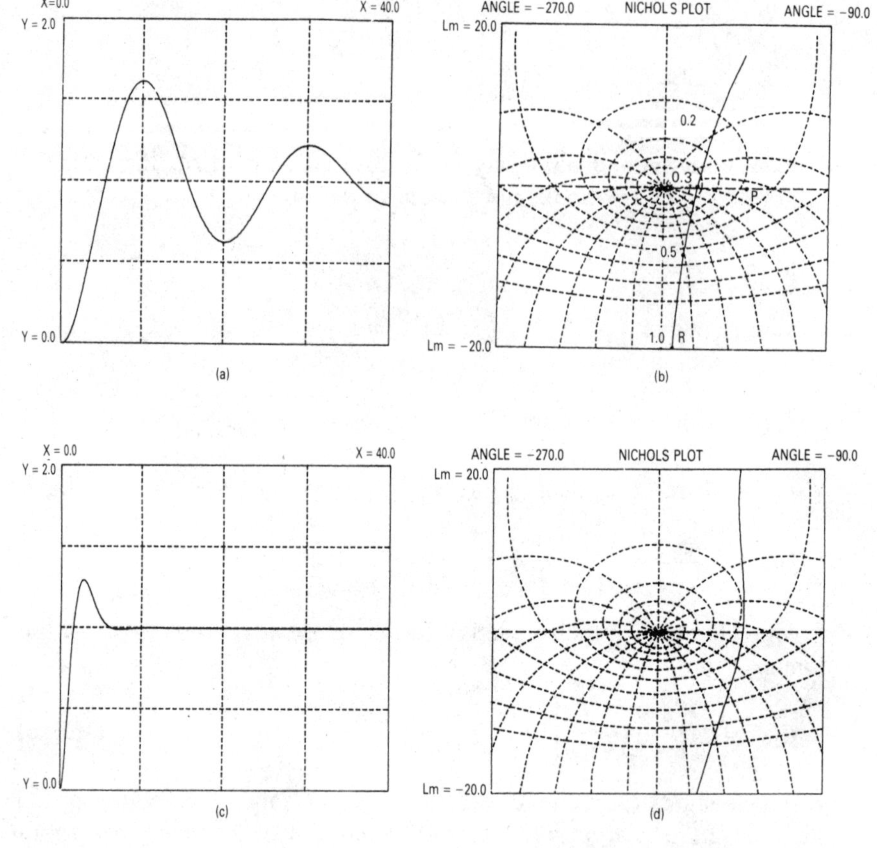

Figure 3.6 (a) Closed loop step response of the uncompensated system.
(b) Frequency response of the uncompensated open loop system.
(c) Closed loop step response of the system with analogue compensation.
(d) Open loop frequency response of the system with analogue compensation

To use the phase advance network effectively three things should be borne in mind:

(a) The maximum phase advance that the network produces is φ_c and is given by

$$\sin \varphi_c = \frac{k - 1}{k + 1} \tag{3.17}$$

where k, the advance ratio, is defined in eqn (3.16).

(b) φ_c occurs at ω_c, the centre frequency of the network, and the time constant

$$T_c = \frac{1}{\omega_c \sqrt{k}} \text{ sec} \tag{3.18}$$

(c) The gain of the network at the centre frequency is

$$M_c = G\sqrt{k} \tag{3.19}$$

Appendix D shows how these expressions are arrived at.

In the present example, the aim is to alter the shape of the original curve so that (see figure 3.6b) the point R is moved to P, creating a phase margin of 45° as required, and at a frequency of 1 rad/sec. It can be seen that this involves a phase change of 39°, which necessitates (eqn 3.17) an advance ratio of $k = 4.4$.

The centre frequency ω_c is chosen to be 1 rad/sec so that, from eqn (3.18)

$$T_c = 1/(4.4)^{1/2} = 0.48 \text{ sec}$$

The necessary vertical movement in the transition from R to P is 20 dB, some of which is provided by the constant G and some by the term

$$\left[\frac{1 + j\omega_c k T_c}{1 + j\omega_c T_c}\right] \tag{3.20}$$

The contribution of the latter is, substituting (3.18) into (3.20), $\sqrt{k} = 2.1$. Since the total movement required is 20 dB, a factor of 10, this must be M_c in eqn (3.19) so

$$2.1\,G = 10 \qquad \text{or} \qquad G = 4.76$$

The compensating network is therefore

$$C(s) = 4.76 \left[\frac{1 + 2.11s}{1 + 0.48s}\right] \tag{3.21}$$

The compensated closed loop step response is shown in figure 3.6c and the corresponding open loop frequency response in figure 3.6d.

Digital design by emulation. $C(s)$ as defined by eqn (3.21) now forms the basis of the emulation and the only question to be settled is the sampling interval, assuming the use of the bilinear transformation. Figure 3.7 shows the closed loop step response of the compensated system for four values of T. The smallest sampling interval, 0.1 sec, is seen to produce a response almost identical to that of the analogue system (figure 3.6c), and the largest, 1.0 sec, is obviously unacceptable. A suitable choice for T would in

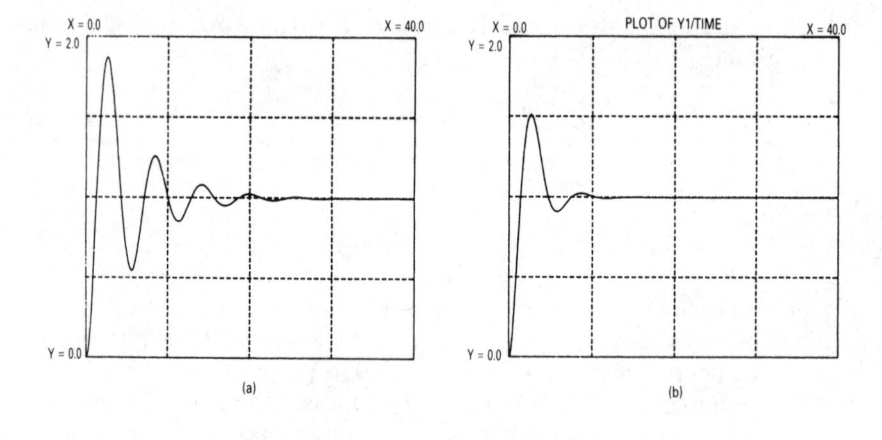

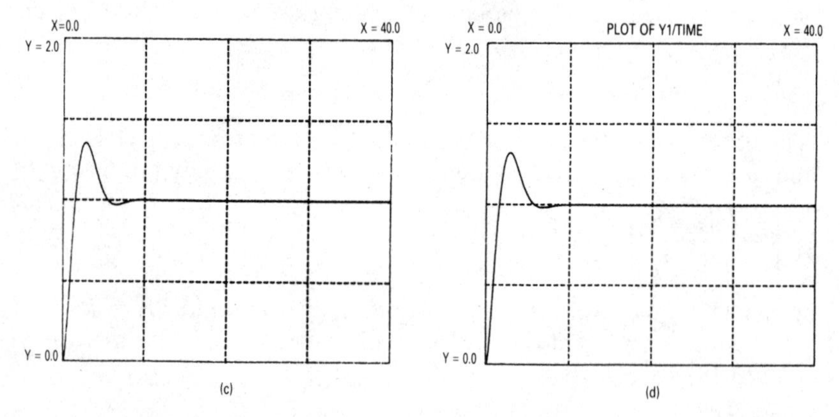

Figure 3.7 (a) Closed loop step response with a digital compensator designed by emulation, using the bilinear transform with $T = 1$ sec.
(b) With $T = 0.5$ sec.
(c) With $T = 0.15$ sec.
(d) With $T = 0.1$ sec

this example probably be in the range 0.1 sec to 0.2 sec. Using $T = 0.15$ sec:

$$C(z) = \frac{18.65405 - 17.36757z^{-1}}{1 - 0.729730z^{-1}}$$

Digital design on the w*-plane*. Since the bilinear transform and many other methods work very well at small values of T, let us examine the w-plane approach for, say, $T = 1$ sec, at which value emulation methods fail:

$$\frac{1}{s}F(s) = \frac{1}{s^2(1 + 10s)} = f_p(s)$$

Consulting the transform table in Appendix A and setting $T = 1$ sec leads to

$$F_z(z) = \frac{z - 1}{z} Z\{f_p(s)\} = \frac{0.048374z^{-1} + 0.046788z^{-2}}{1 - 1.904837z^{-1} + 0.904837z^{-2}}$$

This function is then transferred to the w-plane using the bilinear transform to give

$$F_z(w) = \frac{(1 + 0.0083w)(1 - 0.5w)}{w(1 + 10w)}$$

A plot of this function on the gain-phase plane is shown in figure 3.8a and it is interesting to compare this with figure 3.6b; both are uncompensated frequency response loci and the difference illustrates the reduction in loop stability that is brought about by the mere fact of digitisation with $T = 1$ sec. (More is said about this in Chapter 4.)

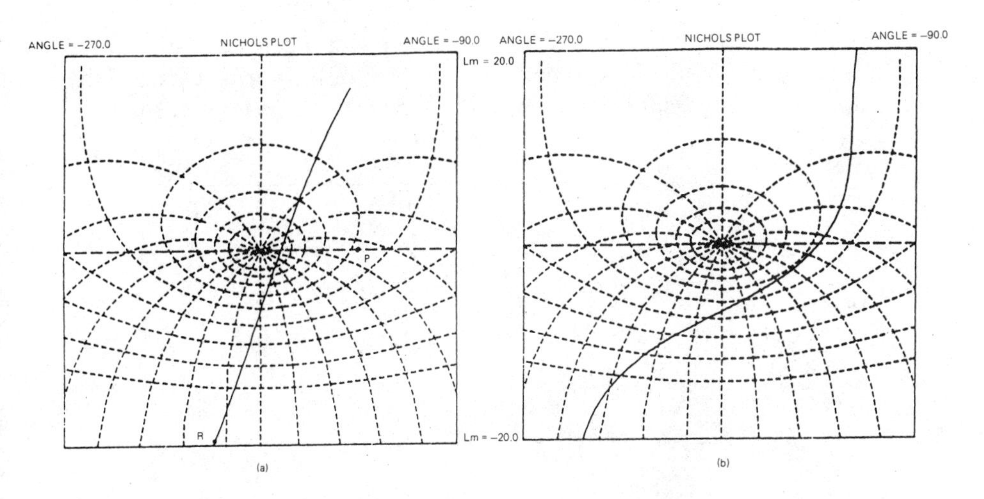

Figure 3.8 (a) Compensation on the w-plane: the frequency response for $F_z(w)$ before compensation, with $T = 1.0$ sec. (b) Compensation on the w-plane: the frequency response for $C(w)F_z(w)$, with $T = 1.0$ sec

The point R, corresponding to $\omega = 1$ rad/sec, is now at (−19.1 dB, −200°) so the value of φ_c is increased considerably over the analogue value:

$$\varphi_c = 200° - 135° = 65°$$

Hence, from eqns (3.17)–(3.19);

$$k = 20$$
$$T_c = 0.224 \text{ sec}$$
$$G = 2.0$$

Thus

$$C(w) = 2.0\left[\frac{1 + 4.47w}{1 + 0.224w}\right] \tag{3.22}$$

The compensated frequency response in the w-plane is shown in figure 3.8b.

The expression for $C(w)$ is now translated back to the z-domain via the bilinear transform, to give

$$C(z) = \frac{13.729 - 10.967z^{-1}}{1 + 0.38122z^{-1}} = \frac{x_f(z)}{x_e(z)} \tag{3.23}$$

Hence the processor software will realise the algorithm

$$x_f(k) = 13.729\, x_e(k) - 10.967\, x_e(k-1) - 0.38122\, x_f(k-1)$$

The closed loop step response of the system with the digital controller is shown in figure 3.9 and it can be seen that most of the performance of the analogue controller has been recovered, despite the long sampling interval.

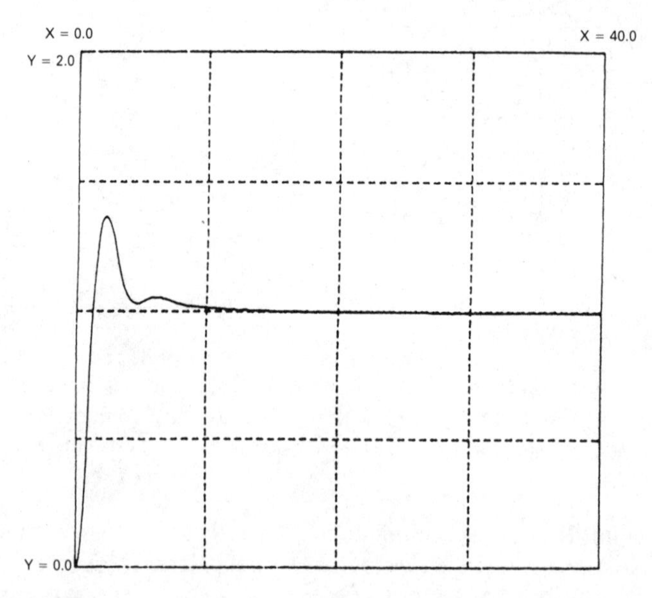

Figure 3.9 Closed loop step response of the system with a digital compensator designed on the w-plane for $T = 1.0$ sec

Figure 3.10 shows the plant input and output during the step response in a little more detail.

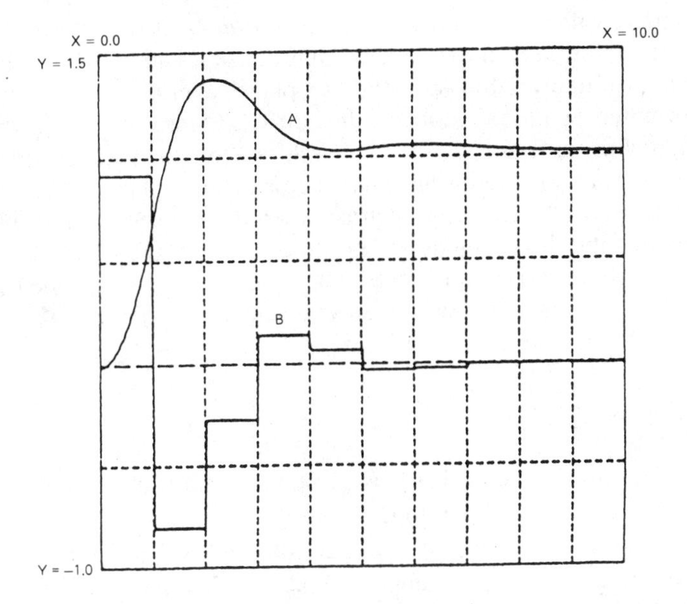

Figure 3.10 Curve A is the same response as that shown in figure 3.9. Curve B is the corresponding output from the controller

3.4 Some properties of the *z*-plane

As stated at the beginning of this chapter, the design methods emphasised in this book are those carried out by emulation and on the *w*-plane. It is wise however to acquire some understanding of the *z*-plane and its basic properties, and for this reason we have included the following section. (Reference [Og87] is commended to those who wish to pursue *z*-plane design techniques fully.) The poles and zeros of discrete transfer functions can be plotted on an Argand diagram to form the *z*-plane, just as continuous transfer functions can be represented on the *s*-plane. From this, it is possible to develop design methods for discrete controllers which are analogous to the classical root locus methods on the *s*-plane. Although we have chosen not to cover *z*-plane root locus design methods, a discussion of the principal features of *z*-plane representation is nevertheless useful; the main concepts are straightforward. Any discrete transfer function can be represented by an expression of the form

$$F(z) = K\frac{(z - z_1)(z - z_2)(z - z_3)\ldots(z - z_n)}{(z - p_1)(z - p_2)(z - p_3)\ldots(z - p_m)}$$

in which z_1, z_2, etc. are the zeros of the transfer function and p_1, p_2, etc. are the poles, some of which may of course occur in complex conjugate pairs.

A prime concern is the condition for stability, which in the s-domain requires poles with negative real parts. In section 2.4 it was shown that the equivalent condition in the z-plane is that the poles should lie within the unit circle (see figure 2.8). As with the s-plane, overall system stability is not determined by the positions of the zeros, so these may lie anywhere, but a plant with zeros outside the unit circle is liable to present difficulty in compensation (as with right half plane zeros in the s-plane).

One of the most fruitful ideas to grasp when thinking about the z-plane is that of evaluating the frequency response of a digital filter from a z-plane plot of the filter's poles and zeros. (It is an equally useful idea in the s-plane, where it may be stated in an equivalent form.)

Consider the simple example

$$G(z) = K\left[\frac{z - b}{z - a}\right]$$

To determine the frequency response, one must substitute $j\omega$ for s, that is $e^{j\omega T}$ for z; it is then necessary to evaluate the expressions $(e^{j\omega T} - a)$ and $(e^{j\omega T} - b)$. Since $e^{j\omega T} = \cos \omega t + j\sin \omega t$, the term $e^{j\omega T}$ may be described as a vector of unit length at an angle ωT and may therefore be drawn on the z-plane as OP in figure 3.11. Since a and b may be represented by OA and OB, then

$$\text{OP} - \text{OA} = \text{AP}$$

and

$$\text{OP} - \text{OB} = \text{BP}$$

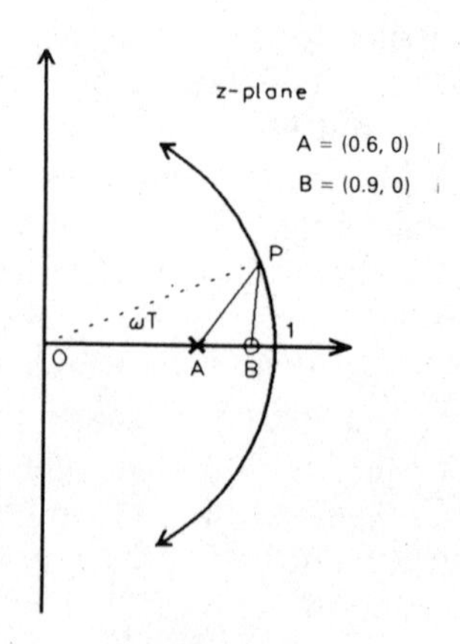

Figure 3.11 The pole/zero constellation for a simple digital filter, a phase advance network

$G(z)$ evaluated for $z = e^{j\omega T}$ is therefore seen to be

$$G(j\omega T) = K(\text{BP/AP})$$

For a given sampling interval T, the point P is determined by the value chosen for ω, so we can, for that sampling frequency, mark the unit circle off in values of signal frequency. It can be seen that the point $z = 1$ corresponds to $\omega = 0$ and $z = -1$ to $\omega = \pi/T = \frac{1}{2}(2\pi/T) = \frac{1}{2}f_s$ – the highest frequency that can be sampled without aliasing, the Nyquist limit.

Each factor in a discrete transfer function, such as $F(z)$ above, contributes to the overall frequency response in the same way as that just explained for $G(z)$. Clearly, the magnitude of $F(j\omega T)$ is given by

$$|F(j\omega T)| = K\frac{|z - z_1||z - z_2| \ldots |z - z_n|}{|z - p_1||z - p_2| \ldots |z - p_m|}$$

and the phase by

$$\angle F(j\omega T) = \angle(z - z_1) + \angle(z - z_2) + \ldots + \angle(z - z_n)$$
$$-\angle(z - p_1) - \angle(z - p_2) - \ldots - \angle(z - p_m)$$

Since $|z - z_1|$ and $\angle(z - z_1)$ are represented diagrammatically by the vector from z_1 to z, it is possible to assess a transfer function's frequency response from the positions of its poles and zeros on the z-plane, with respect to a point moving around the unit circle as ω increases. The following examples illustrate the process.

Example 3.2

(a) A pole at $z = 0.6$ and a zero at $z = 0.9$, as shown in Figure 3.11, corresponds to the discrete transfer function

$$F_a(z) = K\left(\frac{z - 0.9}{z - 0.6}\right)$$

At zero frequency, there is a gain of $0.25K$. As ω increases, the vectors from the unit circle to the pole and zero become similar in length, indicating a gain tending towards K, although as ωT reaches $\pi/2$ and beyond it will slightly exceed this. The phase, indicated in this case by the angle APB, see figure 3.11, starts at zero, increases in a positive direction and then falls back towards zero. Geometrical analysis could obviously be used to determine the maximum phase and the frequency at which it occurs. Increasing gain and leading phase over a range of frequencies is characteristic of a phase-advance compensator; this therefore is a discrete form of the phase advance network used in Example 3.1, section 3.3.

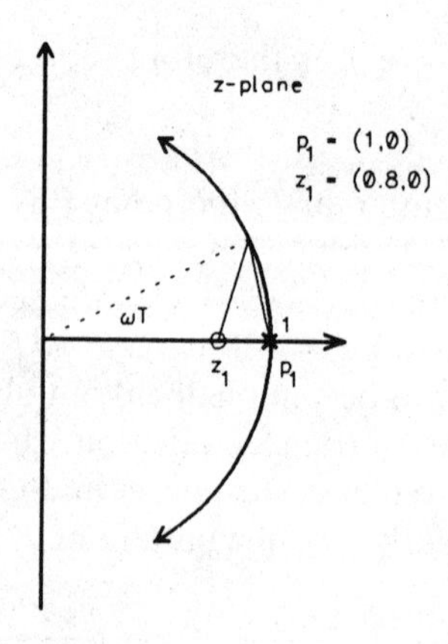

Figure 3.12 The pole/zero constellation for a digital form of PI network

(b) A pole at $z = 1$ and a zero at $z = 0.8$, as shown in Figure 3.12, have a corresponding discrete transfer function:

$$F_b(z) = K\left[\frac{z - 0.8}{z - 1}\right]$$

At zero frequency, the length of the vector to the pole is itself zero, indicating an infinite gain; as the frequency increases, its length becomes non-zero giving a falling gain, and for ωT greater than about 30° the lines to the zero and the pole again become similar in length, meaning that the gain approaches unity. The phase at $\omega = 0$ is not meaningful, but for very low frequencies can be seen to be close to −90°, progressively changing towards zero as ωT increases. This characteristic is typical of a Proportional plus Integral or PI compensator. This, then, is a digital form of PI network.

It is hoped that these two examples have given an appreciation of some of the properties of the z-plane. The principles illustrated are just the same with complex poles and zeros, although obviously the interpretation is more complicated. A further example (Example 5.3) is included in Chapter 5 which has zeros on the unit circle, although its design is not presented in detail.

3.5 The frequency response of sampled data systems

We have discussed the concept of frequency response as applied to discrete networks and it is assumed that the reader is familiar with its application to continuous systems. How then do we deal with sampled data networks, some components of the system being analogue and some digital?

The approach that is most often adopted is to transfer the *z*-domain blocks into the *s*-domain by emulation; if the sampling frequency is reasonably high, this is likely to work well enough for most purposes.

To do the job 'properly', however, we must undertake a full analysis of the system, as explained in Chapter 2, and extract the frequency response from that. To understand the procedure, consider again Example 2.7 in the previous chapter; the topology is shown in figure 2.20 and reproduced below as figure 3.13. (Remember that a *z*-block is considered to have a sampler at its input and a ZOH at its output.)

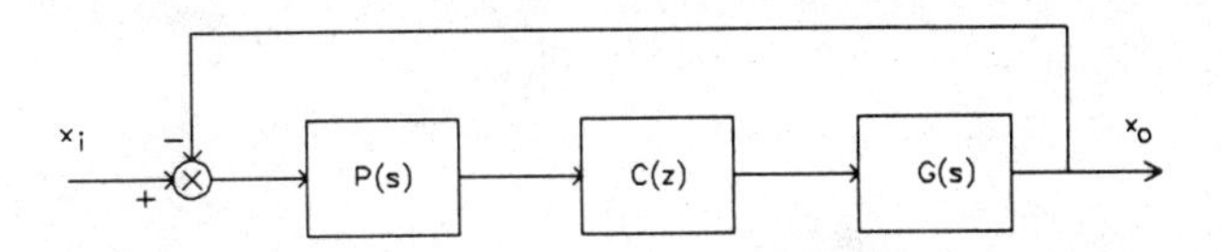

Figure 3.13 A sampled data system for which the frequency response is to be determined

From the analysis in Chapter 2:

$$x_o(z) = \frac{(X_iP)CG_z}{1 + C(GP)_z} \tag{3.24}$$

where

$$(X_iP) \triangleq \mathfrak{Z}\,\{x_i(s)P(s)\}$$

$$G_z \triangleq (1 - z^{-1})\,\mathfrak{Z}\left\{\frac{1}{s}\,G(s)\right\}$$

$$(GP)_z \triangleq (1 - z^{-1})\,\mathfrak{Z}\left\{\frac{1}{s}\,G(s)P(s)\right\}$$

Since we are considering the frequency response, it follows that $x_i(t)$ is a sinewave. The product $x_i(s)P(s)$ may be expanded into partial fractions, some of which will be found to relate directly to the sinewave while others are related to the poles of $P(s)$. It follows that the expression for $x_o(z)$ as defined in eqn (3.24) may be re-written as

$$x_o(z) = \frac{CG_z}{1 + C(GP)_z}\,[\mathfrak{Z}\,\{x'(s)\} + \mathfrak{Z}\,\{x''(s)\}] \tag{3.25}$$

where $\mathscr{Z}\{x'(s)\}$ contains the terms that relate to the input sinusoid. The remaining terms may therefore be discarded to leave us only the frequency information.

Since $\mathscr{Z}(x'(s)\}$ is simply the z-transform of the sinusoid emerging from the block $P(s)$, it can be seen that the effect of changing eqn (3.24) to (3.25) and discarding $\mathscr{Z}\{x''(s)\}$ is to change figure 3.13 to figure 3.14. In eqn (3.25), the factor

$$\frac{CG_z}{1 + C(GP)_z} = F(z) \tag{3.26}$$

is the closed loop transfer function linking x_i' to x_o in figure 3.14, and this leads directly to the frequency response of that section of the figure, by substituting $e^{j\omega T}$ for z to give $F(j\omega T)$. The complete frequency response is then, from fig 3.14:

$$H(j\omega) = P(j\omega)\,F(j\omega T) \tag{3.27}$$

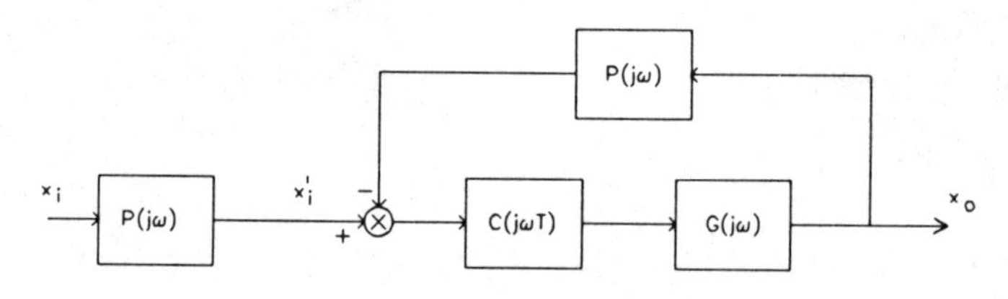

Figure 3.14 A sampled data system with a frequency response equivalent to that of the network in fig. 3.13

An alternative approximation to this procedure would be to express all the blocks of the system in frequency response form ($C(j\omega T)$, $G(j\omega)$, $P(j\omega)$) and to compute the overall closed loop frequency response, from figure 3.13, as

$$H(j\omega) = \frac{P(j\omega)C(j\omega T)G(j\omega)}{1 + P(j\omega)C(j\omega T)G(j\omega)} \tag{3.28}$$

The earlier method of approximation, whereby $C(z)$ is transformed into $C(s)$ by some technique of emulation, would obviously lead to

$$H(j\omega) = \frac{(Pj\omega)C(j\omega)G(j\omega)}{1 + P(j\omega)C(j\omega)G(j\omega)} \tag{3.29}$$

Clearly, the values of $H(j\omega)$ computed by each of these three methods (eqns (3.27)–(3.29)) would differ, and the only one that is strictly correct is (3.27). The computation of $F(z)$ in eqn (3.26) is far from simple however, even when $P(s)$, $C(z)$ and $G(s)$ are trivial; in practice, one could only consider undertaking such an exercise with the aid of well-written software. In many cases, therefore, the use of one of the approximations is the only realistic course.

If the software is available to carry out the complex computations involved, then more complicated examples than the one just cited can be undertaken in the same way. A case in point could be Example 2.9, figure 2.22, reproduced as figure 3.15; it is shown in Chapter 2 that

$$x_o(z) = (X_iG) - \frac{(X_iGQ)PG_z}{1 + P(GQ)_z}$$

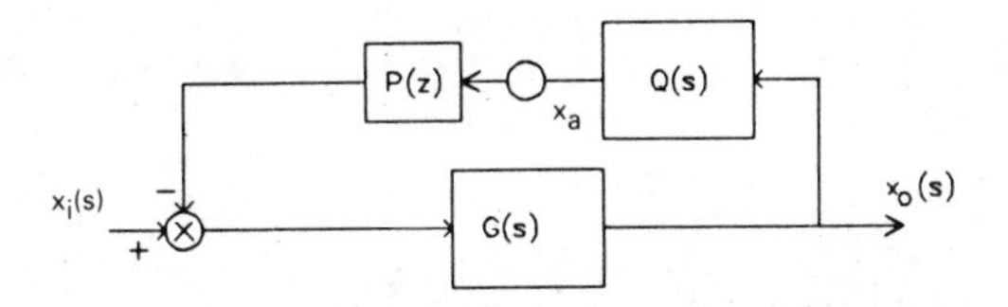

Figure 3.15 Another example of a sampled data system whose frequency response is required

Consider (X_iG): this means, in words, 'the z-transform of the signal emerging from $G(s)$'. If we discard all but the sinusoidal component, we are taking the z-transform of the sinewave emerging from $G(s)$, described in magnitude and phase by $G(j\omega)$.

Likewise for (X_iGQ), which by the same reasoning leads to the product $G(j\omega)Q(j\omega)$. If we denote

$$F(z) = \frac{PG_z}{1 + P(GQ)_z}$$

then $F(j\omega T)$ denotes this expression with $z = e^{j\omega T}$, and the frequency response of the system in figure 3.15 is given by

$$H(j\omega) = G(j\omega)(1 - Q(j\omega)F(j\omega T))$$

3.6 Compensation of a complex sampled data system

Compensator design was earlier considered at some length but only in the case where the topology was simple – merely the compensator followed by the plant.

What if, to increase the complexity only slightly, the system were that of figure 3.16?

Consider firstly the problem of design in the s-domain, leading of course to z-plane design by emulation.

Most of the simple classical design techniques are intended for unity feedback systems so the presence of $F(s)$ in figure 3.16 precludes their use, at least directly. To make use of frequency response methods, such as the

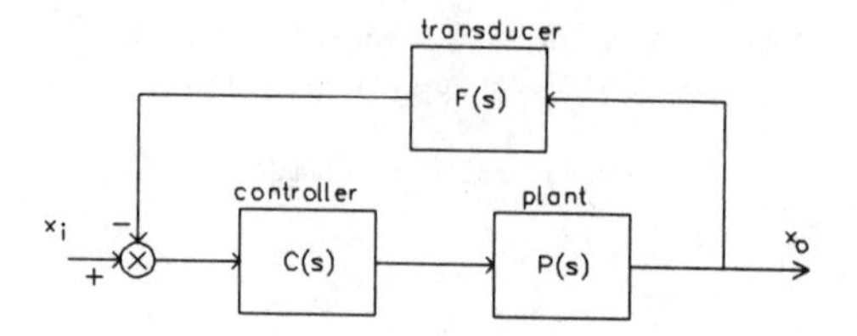

Figure 3.16 An example of a feedback system that does not have unity feedback

Nichols chart, we re-cast the system into the standard form shown in figure 3.17. The input/output relationship for both figures is the same, so we may write

$$\frac{D(s)G(s)}{1 + D(s)G(s)} = \frac{C(s)P(s)}{1 + C(s)P(s)F(s)} \tag{3.30}$$

If now we suppose that in both cases the controller is absent, that is, $D(s) = C(s) = 1$, it can be shown from eqn (3.30) that

$$G(s) = \frac{P(s)}{1 + P(s)[F(s) - 1]} \tag{3.31}$$

When the controllers are included however we may deduce, again from eqn (3.30), that

$$C(s) = \frac{D(s)}{1 - P(s)[F(s) - 1][D(s) - 1]} \tag{3.32}$$

The significance of this is that the unity feedback, single loop, system of figure 3.17 can be designed in the classical manner, knowing $G(s)$ from eqn (3.31); this gives us $D(s)$, which gives $C(s)$ from eqn (3.32).

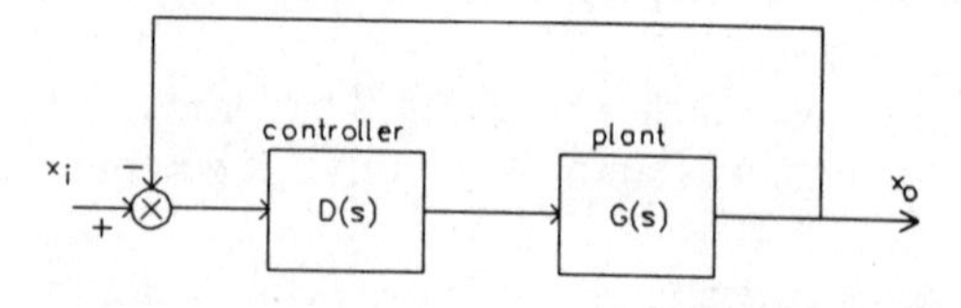

Figure 3.17 The standard form of unity feedback system

More complex systems can be treated in the same way, for example that of figure 3.18, the closed loop transfer function of which is equated to that of its single loop unity feedback equivalent, to give (omitting (s) for convenience)

$$\frac{DG}{1 + DG} = \frac{CP}{1 + CF + CPH} \tag{3.33}$$

As before, setting $D = C = 1$ we define

$$G = \frac{CP}{1 + F + P(H - 1)} \tag{3.34}$$

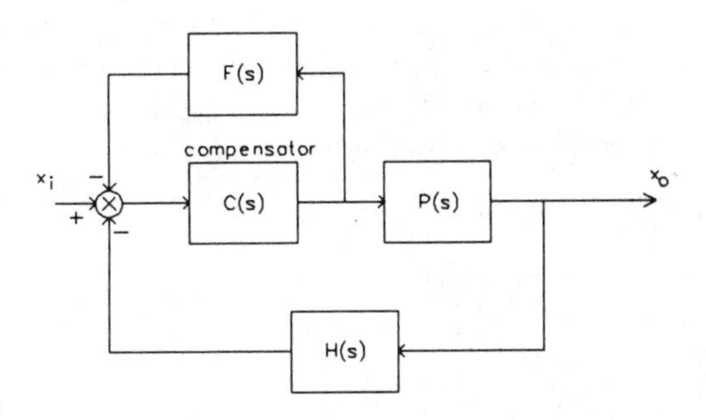

Figure 3.18 A more complex system for compensator design

Now including the compensators, C is defined in terms of D from eqn (3.33) as

$$C = \frac{DG}{P(1 - DG(H - 1)) - DGF} \tag{3.35}$$

If this is put into practice without the aid of appropriate software, the technique would be tiresome, to say the least – but the software required is easily written. There is of course no guarantee that the resulting controller is realisable, and certain obvious questions then arise: what does one do if it cannot be realised? How might a complex controller, arrived at in this way, be replaced by a simpler approximation? No solutions have yet been suggested.

A design approach based on the z-transform and the w-plane is little different in concept from that just explained, although the arithmetic demands are even greater. To examine this, let us look again at the previous example but with the controller realised in digital terms as $C(z)$, and as $D(z)$ in the single loop equivalent. The system is shown in figure 3.19.

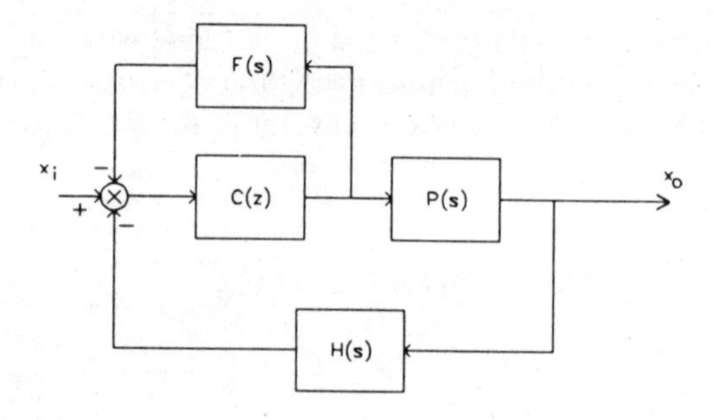

Figure 3.19 The system of figure 3.18 with digital compensation

Since all functions in the following equations are expressed in the z-domain, we omit the argument (z).

Conducting the analysis as explained in Chapter 2:

$$x_o = x_a C P_z$$

$$x_a = x_i - x_a C F_z - x_a C(PH)_z$$

that is

$$x_a = \frac{x_i}{1 + C(F_z + (PH)_z)}$$

$$\therefore \quad \frac{x_o}{x_i} = \frac{CP_z}{1 + C(F_z + (PH)_z)} \tag{3.36}$$

Of the equivalent circuit we can write

$$\frac{x_o}{x_i} = \frac{DG_z}{1 + DG_z} \tag{3.37}$$

Setting $D = C = 1$ and equating the transfer functions in eqns (3.36) and (3.37) yields

$$G_z = \frac{P_z}{1 + F_z + (PH)_z - P_z G_z}$$

$D(z)$ may now be designed given this expression for G_z, and thus $C(z)$ is determined by equating the transfer functions again. This leads to the lengthy expression

$$C(z) = \frac{DG_z}{P_z(1 + DG_z) - DG_z(F_z + (PH)_z)}$$

3.7 Multiple sampling rates

In practice, most sampled data systems are simple in their topology but many will have more than one loop; a typical example is shown in figure 3.20.

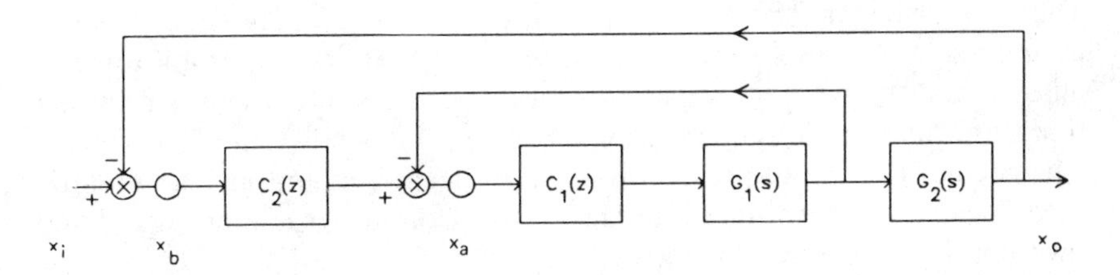

Figure 3.20 A feedback system with digital controllers using different sampling rates

By the principles given in Chapter 2, the input/output equations are

$$\begin{aligned} x_o &= x_a C_1 (G_1 G_2)_z \\ x_a &= x_b C_2 - x_a C_1 (G_1)_z \\ x_b &= x_i - x_a C_1 (G_1 G_2)_z \end{aligned} \tag{3.38}$$

and lead to the transfer function

$$\frac{x_o}{x_i} = \frac{C_1 C_2 (G_1 G_2)_z}{1 + C_1 (G_1)_z + C_1 C_2 (G_1 G_2)_z} = F(z)$$

The inner loop in such a system may well enclose a high-speed, fast-acting portion of the plant and thus require a high sampling rate in $C_1(z)$; the outer loop, on the other hand, may be appreciably slower, depending on the nature of $G_2(s)$, so that the sampling rate for $C_2(z)$ can be relatively slow. There is nothing that requires both digital controllers to sample at the same rate [As84] and no great complexity is introduced in practice by a difference. More will be found on this in Chapter 6.

On the theoretical side however, it is necessary to exercise just a little care. Some of the factors in the expression for $F(z)$ must be evaluated at one sampling rate and some at the other; consideration of eqns (3.38) describing the system should clarify which is which. $(G_1)_z$ and $(G_1G_2)_z$ are both derived from responses to a pulse at the higher sampling rate and so both are evaluated accordingly. Naturally $C_1(z)$ is evaluated at the higher rate and $C_2(z)$ at the lower. Thus there is little complexity introduced, even in theory, by the dual sampling rate – just the need for care – and however complicated the topology, reasoning like that used above will easily sort out the situation.

3.8 Summary

The chapter began with a discussion of the concept of frequency response as applied to a digital algorithm. It dwelt a little on the important topic of aliasing and the need in many cases for an antialias filter. This is required to ensure that signal frequencies much above the bandwidth of the system (that is, noise) are not present to any significant degree; one can then be sure that the sample values recorded derive from the true input signal and owe nothing to extraneous high-frequency pick-up.

The mathematical argument here is sometimes misconstrued. It does not say that the sampler *creates* high-frequency signals that are found above the baseband, but points out that if such signals exist they will give rise to samples indistinguishable from those generated within the baseband. If the sampler did indeed create high-frequency signals, then a low-pass filter in front of it could not remove them.

Several methods of emulation were described, enabling one to map a transfer function, with varying degrees of success, between the s- and z-domains. The usual methods of doing this were given (the bilinear transform with and without pre-warping, pole and zero mapping, etc.) but a less common approach (described in more detail in Appendix E) is also included which promises to be more accurate than the others although quite simple to apply.

However, digital design in the w-plane has proved in recent years to be probably the most popular and most successful method of designing digital controllers that will operate acceptably at very low sampling rates (which emulation methods will not) and this was described in some detail. Compensator design directly on the z-plane is favoured by some people, but it is not a technique that would in our view be as quickly picked up as those we have described, so it has largely been ignored here.

The frequency response of sampled data networks in general is a topic that is, for some reason, not often discussed in textbooks, although it is obviously relevant. The topic has been introduced here and discussed briefly, together with the associated question of a generalised design technique for sampled data systems. Neither subject is taken far, as we would quickly exceed the scope of this text, but the reader should gain an appreciation of the problems involved and how they may be solved. The rapid spread of digital, and particularly distributed, control makes this a subject of some importance.

Problems

The reader may find Appendix D useful in dealing with problems 1 to 7, which involve the design of phase advance or PI controllers.

1. The angular position of the output shaft is related to the input voltage of a d.c. motor by the transfer function

$$G(s) = \frac{5}{s(s + 1)(0.1s + 1)}$$

A digital compensator is to be designed so that when the motor is controlled within a feedback loop by the compensator in cascade, the system has an adequate phase margin.
The frequency response of the w-plane model of the motor plus ZOH, $G_z(j\omega)$, is tabulated in table Q1

Table Q1 Frequency response of $G_z(w)$

ω *(rad/sec)*	*Gain (dB)*	*Phase (deg.)*
1.0	10.8	−142
1.5	5.3	−156
2.0	0.8	−166
2.5	−2.8	−174
3.0	−5.9	−181
3.5	−8.6	−186
4.0	−11.0	−191
4.5	−13.1	−195
5.0	−15.1	−199

(a) Use $G_z(j\omega)$ to design a phase advance compensator giving the system a phase margin of 45° with as high a crossover frequency as possible. The compensator transfer function when expressed in terms of w should have the form

$$C(w) = K\left[\frac{1 + 10Tw}{1 + Tw}\right]$$

(b) Express the action of the compensator in terms of the time domain equation that the digital processor must implement. The sampling interval is 0.03 sec.

2. A plant described by the transfer function

$$F(s) = \frac{900}{s(s + 30)}$$

is to be controlled within a feedback loop by a digital processor with a sample interval T of 40 msec, that is, 0.04 sec.
(a) Determine the z-domain description of the plant $F_z(z)$, stating clearly any assumptions you think necessary.

(b) The w-plane description is given by

$$F_z(w) = \frac{900(1 - wT/2)}{w(30 + w)}$$

the frequency response of which is tabulated in table Q2.

Table Q2 Frequency response of $F_z(w)$

rad/sec	*dB*	*deg.*
10	9.0	−120
15	5.3	−134
25	0.3	−156
35	−3.4	−174
45	−6.0	−188
60	−9.1	−204

Design a digital controller to create in the compensated system a closed loop peak gain of about 2 dB at a frequency of 45 rad/sec. The controller you design should be described by a transfer function in the z-domain.

3. (a) A PI controller takes the form in the s-domain:

$$C(s) = K\left[\frac{1 + sT_c}{sT_c}\right]$$

Using (i) pole mapping (Appendix E) and (ii) pole and zero mapping, obtain an expression for the controller in digital form. Express your answers as transfer functions in the form

$$C(z) = \frac{a_0 + a_1 z^{-1}}{1 + b_1 z^{-1}}$$

with the coefficients given in terms of T, K and T_c.

(b) A plant described in the s-domain by

$$F(s) = \frac{3}{s^2 + 4s + 3}$$

is to be controlled by a PI compensator in digital form, and since the sampling frequency is low, at 3 Hz, the design should be carried out on the w-plane. Obtain a suitable compensator $C(z)$ and, if a computer is available with the appropriate software, compare the performances of the three controllers you have designed by forming an s-plane controller and emulating as in part (a). The compensated system should have a crossover frequency of 1.5 rad/sec with a phase margin of 45°.

4. A plant described by the transfer function

$$P(s) = \frac{K_p e^{-sTd}}{1 + sTp}$$

is controlled within a unit feedback loop by a compensator

$$C(s) = K\left[\frac{1 + sTc}{sTc}\right]$$

(a) If $K_p = 1.0$
$T_p = 5.0$ sec
$T_d = 2.0$ sec
determine suitable values for K and T_c, when the aim is to create a crossover frequency of 0.3 rad/sec with a phase margin around 55°.

(b) Form a discrete model of the plant $P_z(w)$ and hence design a digital compensator of the form above to meet the same specification when the sample interval is 0.5 sec.

(c) Compare the closed loop step responses of (a) and (b).

(d) Repeat steps (b) and (c) for $T = 1.0$ sec.

5. A digital controller $C(z)$ is used in cascade with a plant $P(s)$ to provide compensation designed to ensure that the closed loop performance of the system, with unity feedback, is both fast and adequately damped. Given that

$$P(s) = \frac{1}{1 + 10s}$$

(a) determine the plant model in the w-plane, $P_z(w)$, for a sampling interval T of 1 sec, assuming the presence of a ZOH as usual.

(b) If $C(z)$ is merely a constant, K, determine the value it must have if the phase margin is to be 70°.

(c) If $C(z)$ is a PI controller, determine the necessary parameter values to give the closed loop system a peak gain of 3 dB at a frequency of 1 rad/sec.

6. To improve the performance of a plant $G(s)$ a digital form of phase advance is added in cascade together with the usual analogue/digital converters and ZOH. The plant transfer function in the w-plane is given by

$$G_z(w) = \frac{\left(1 - \dfrac{w}{4}\right)\left(1 + \dfrac{w}{72}\right)}{w\left(1 + \dfrac{w}{0.66}\right)}$$

when the sampling interval T is 0.5 sec.
The compensator is described by

$$C(w) = K\frac{(1 + 15\tau w)}{1 + \tau w}$$

Determine the values of K and τ required in the compensator to give a phase margin of 55° with as high a crossover frequency as possible. Give the time domain relationship between the compensator input and output quantities.

7. A plant described by

$$G(s) = \frac{12.5}{s(1 + s/1.25)}$$

has an inadequate phase margin of 15°–20° which must be improved to about 50° by a simple cascade compensator that is to be implemented on a digital processor. The compensator should at the same time increase the speed of the response by increasing the crossover frequency by a factor of 2.

Using design on the w-plane produce a suitable compensator, expressing the compensator transfer function in terms of z, given that the plant transfer function in the w-plane is

$$G_z(w) = \frac{12.5(1 - w/20)}{w(1 + w/1.25)}$$

with $T = 0.1$ sec.

8. A simple low-pass filter

$$F(s) = \frac{4}{(s + 2)^2}$$

is to be realised in digital form using a short sampling interval of 1/20 sec.
Produce a digital filter $F(z)$ suitable for the purpose using:

(a) the bilinear transform
(b) the bilinear transform with pre-warping
(c) pole mapping with the Taylor expansion (Appendix E)
(d) the three-point Adams–Bashforth algorithm.

If suitable software is available, compare the frequency responses.

9. One way of determining the position of a freely moving body is to perform a double integration on the outputs of accelerometers attached to it.
Determine the coefficients of a three-point algorithm to do this digitally, given that the sampling period is 1 sec (the body is large and slow moving) and that computation time is negligible by comparison. Use the pole mapping technique described in Appendix E, where the function to be mapped is $1/s^2$.

10. Counting pulses from a rotating disc is a common method of determining position; to compute velocity and acceleration from the pulse count we must differentiate. Use the pole mapping method to determine:

(a) a three-point algorithm for the first derivative
(b) a four-point algorithm for the second derivative.

In both cases, assume that the computation requires a full sampling interval so that the quantity being computed is the value of the first or second derivative T seconds after the latest sample.

11. In the sampled data network of figure 3.13:

$$P(s) = \frac{1}{1 + 0.2s}$$

$$C(z) = 18.654 \left[\frac{1 - 0.93103z^{-1}}{1 - 0.72973z^{-1}}\right]$$

$$G(s) = \frac{1}{s(1 + 10s)}$$

(a) Graph the frequency response of the closed loop system.
(b) Compare it to that obtained when $C(z)$ is replaced by $C(s)$ using the bilinear transform, with $T = 0.15$ sec, in which case (see Example 3.1)

$$C(s) = 4.76 \left[\frac{1 + 2.11s}{1 + 0.48s}\right]$$

(c) Compare the results above with those obtained when $T = 1$ sec,

$$C(z) = 13.729 \left[\frac{1 - 0.799z^{-1}}{1 + 0.381z^{-1}}\right]$$

and $C(s)$ is again obtained by emulation with the bilinear transform.

12. In the system depicted in figure 3.20, the controllers $C_1(z)$ and $C_2(z)$ were obtained, using the bilinear transform, from

$$C_1(s) = 4 \left[\frac{1 + 0.016s}{0.016s}\right]$$

and

$$C_2(s) = 5 \left[\frac{1 + 0.016s}{1 + 0.004s}\right]$$

The plant is described by

$$G_1(s) = \frac{20}{s}$$

and

$$G_2(s) = \frac{5}{s(1 + 0.05s)}$$

The inner loop is sampled at 1000 Hz (or $T_1 = 0.001$ sec) and the outer loop at 500 Hz (or $T_2 = 0.002$ sec).
Determine (a) the transfer function $F(z)$ of the overall system, and (b) the sytem frequency response.

4 Error Mechanisms, Filter Structure, and the Sampling Interval

This chapter is different in kind from those that have gone before: much of the material is not to be found in other books, and even research journals are not very helpful. It has nevertheless been felt necessary to include such material in this volume because it is of fundamental concern, and as such should be discussed. This entails introducing a number of topics that are still subject to close research but which can nevertheless be described meaningfully in broad terms. As a consequence, mathematical expressions for certain quantities of interest are quoted here and there without deriving them; the derivations, it is felt, are rather intricate considering the nature and objectives of the book, but the expressions are seen to be quite tractable and lend themselves readily to computer implementation. Appropriate references are given, of course, for those who wish to learn more. Presenting the reader with underived mathematical results is a practice we have adopted as seldom as possible, but it was felt that to present the derivations as well as the results would be a large undertaking of dubious value in the circumstances; we can only hope that our judgement in the matter has been correct.

The chapter begins with a brief description of the types of error that afflict digital filters, and is given in sufficient detail to allow a discussion of the importance of filter structure. The discussion is centred on three particular structures, two of them commonly used (though one of them perhaps ought not to be); the third is seldom encountered, as yet, but deserves a wider popularity.

Filter error is then examined in more detail, to discover how its various manifestations change with the sampling frequency. It turns out that some errors increase while others decrease as the sampling frequency is raised, and there is a certain dependence upon structure.

However, the choice of T affects more than error generation; the mere fact of sampling is of itself the cause of a certain reduction in loop stability, and this is suggested as a limiting factor in choosing an upper value for T (a lower limit for the sampling rate). The sampling interval is limited at the other end of its range, too. Because of the finite length of the computer word, higher sample rates can only be achieved with increasing precision of computation, and there comes a point at which this can no longer be justified.

The chapter concludes with a discussion of how one might in practice make a choice among system parameters such as processor wordlength, controller algorithm, sampling interval, and A/D converter length. Such a choice must be consistent: the errors contributed by the various sources should be in roughly equal proportions.

4.1 Errors in digital filters

For a proper discussion of both filter structure and sampling frequency, it is necessary to have an understanding of the mechanisms by which errors in digital filters arise. This leads directly to the need for a clarification of just what is meant by the term 'error' in this context. When the discrete algorithm is derived by emulating a continuous one it is easy to define the error as the difference between the two, and that is the meaning adopted here. It is more difficult to arrive at a satisfactory definition in the case of an algorithm designed directly in the z-domain. We suggest that the error be defined in that case with respect to the continuous function derived from the discrete one through the bilinear transform, though of course the definition is arbitrary. Since the aim, as we shall see, is to discover what causes errors to increase or decrease, it is not likely that the choice of definition for the measurement datum will be important.

Fundamentally all errors are caused by the quantisation, in both time and amplitude, that is inherent in the nature of digital filters. However the term 'quantisation error' is reserved by convention for those errors that are due purely to quantisation in amplitude; quantisation in time is the source of what are variously called 'truncation' or 'discretisation' or 'algorithmic' errors. Since the term 'truncation error' is easily misconstrued, the choice in this text lies between the other two, and 'algorithmic error' is preferred simply on grounds of euphony.

Thus when writing about errors, it is convenient to divide them at the outset into two types – algorithmic and quantisation, e_a and e_q respectively. Since these are distinct and uncorrelated, the total error is taken to be

$$e_t = e_a + e_q$$

4.1.1 *Algorithmic error*

For a given filter, the algorithmic error e_a depends only upon the sampling interval and the rates of change in the input signal and may be written

$$e_a = J_0 u + J_1(T) \overset{(1)}{u} + J_2(T^2/2!) \overset{(2)}{u} + \ldots + J_n(T^n/n!) \overset{(n)}{u} + \ldots$$

where $\overset{(n)}{u}$ is the nth derivative of the input u. When the digital filter and its analogue counterpart have the same zero-frequency gain, which is normally the case, then $J_0 = 0$. Evaluation of the other coefficients is discussed in section 4.3.

If the computer wordlength and that of the ADC were sufficiently long that quantisation effects could be discounted, algorithmic error would still remain and would account for the difference in performance of equivalent digital and analogue filters. The nature of algorithmic error may perhaps be appreciated more readily by considering, for example, the bilinear transformation. Since this means no more than trapezoid summation (section 3.2.1) it is clear that any *linear* input will be handled without error; there is no error, in other words, proportional to the first derivative. If on the other hand the method of integration were merely rectangle summation, there *would* be a component of error due to the first derivative; conversely if the integration were to be carried out by a three-point Adams–Bashforth algorithm (see section 3.2.2), there would be no error due to either first or second derivatives.

When the means of emulation is based on some numerical integration method, the dependence of the error on the derivatives of the input is reasonably clear, but what is the position when, say, pole and zero mapping is used to generate the function of z? This is a much more complex question but it has been shown [Fo86] that the errors in that case are due to *all* derivatives including the first (which goes some way towards explaining the indifferent performance attributed to this form of emulation).

To arrive at a figure for algorithmic error in any particular case, it is necessary to consider the form and numerical value of the input $u(t)$. A suitable choice for this could be a sinewave of the natural frequency of the system, of an amplitude that is considered, bearing in mind the circumstances, to be neither too large nor too small. Figures for the maximum values of the various derivatives therefore become available and so an estimate of the maximum value of e_a can be arrived at. One must be careful in such an exercise not to create a worst-case estimate that is too pessimistic to be useful; a good deal of work is yet to be done in this area.

4.1.2 Quantisation error

The quantisation error e_q has three components (see [Fo89]):

(1) e_s, the error introduced in the sampling process: referred to as *sampler quantisation error*.
(2) e_r, the error introduced by truncating a multiple wordlength product to some shorter length during arithmetic operations: this is unavoidable

at some point in a recursive filter (one with feedback) and is referred to as *multiple word truncation error*. It has often been given the name 'round-off error', but this is too easily confused with e_s or e_c, and is rejected here for that reason.

(3) e_c, the error caused by representing the filter coefficients to a finite number of bits: referred to as *coefficient representational error*.

These error types will be dealt with again in more detail in section 4.3; general descriptions are given now so that the topic of filter structure can be introduced.

Sampler quantisation error, e_s

This can be regarded as a random noise signal superimposed upon the input; clearly it is subject to the same transfer function as the input itself. On the basis of some reasonable assumptions about the statistical nature of the input noise, one can compute the mean square value of the output noise. It can only be reduced by increasing the number of bits in the ADC but, in a control context, e_s is not usually a source of difficulty. The range of ADCs normally available (8 to 12 bits) should prove sufficient.

Multiple word truncation error, e_r

The product of two numbers of length n bits and m bits has length $(n + m)$ bits. Thus the product of an n-bit coefficient, say b_1, and the previous output value $y(k - 1)$ held to m bits, will be longer than either, and must at some point be reduced to m bits before storage in memory. The discarded bits represent the error e_r.

If the arithmetic operations are arranged carefully to minimise e_r, then this effect will be negligible. (Note the caveat, however, in section 4.3.3.) There are considerable problems inherent in the design of the internal arithmetic of a digital filter; the optimum arrangement is by no means obvious, and the problems are discussed in detail in Chapter 5.

Coefficient representational error, e_c

Coefficient representation is by far the most potent source of quantisation error. The smallest number of binary digits likely to form a word is eight and if the coefficient value is reasonably large, between 0.5 and 1.0, say, the error in representation will be of the order of 0.5%. This may seem small but it can lead to large deviations in the filter output because of the nature of the numerical computation. For example, suppose

$$F(s) = \frac{1}{s + 1}$$

is translated into the z-domain via the bilinear transform, to give (with $T = 0.05$ sec)

$$F(z) = \frac{0.024390(1 + z^{-1})}{1 - 0.951220z^{-1}}$$

Expressing the denominator coefficient to 8 bits changes it to -0.94921875, an error of 0.2%. Even if the numerator coefficient were somehow to remain correct, the steady-state error in the filter response would be 4%; recognising that in reality the numerator coefficient is also approximated (to 0.0234375 – an error of 3.9%) the gain of the filter in the steady state is now

$$\frac{(0.0234375)(2)}{1 - 0.94921875} = 0.923$$

Since the correct figure is 1.0, the error is 7.7% at the stated value of T. This effect is examined in detail in Chapter 5.

4.2 The significance of structure

At its most superficial level, structure could be defined as the shape of the flow diagram which, like the transfer function, defines the filter. However a transfer function can in general be represented by several different flow diagrams or structures, although only one has so far been used – the direct form.

The real point about different filter structures is that they correspond to different sequences of arithmetic operations; this would make no difference if it were possible to work with infinite precision, but of necessity the precision is finite. Quantisation errors of various kinds, as explained above, are created at certain points in the arithmetic process, and the order in which operations are defined determines the magnitude of the quantisation error in the outcome [Li71].

There are a surprising number of different ways in which one may present diagrammatically any given transfer function, but it is not the intention here to attempt an exhaustive list of structures. In fact we will discuss only three, though [Ph90], for example, treats the topic at some length. (Even that does not include all the possibilities.)

To illustrate the point, consider three ways of representing

$$F(z) = \frac{a_0 + a_1z^{-1} + a_2z^{-2}}{1 + b_1z^{-1} + b_2z^{-2}} = \frac{y(z)}{u(z)} \tag{4.1}$$

(1) The simplest is the direct form, introduced in Chapter 2, which leads to the flow diagram of figure 4.1.

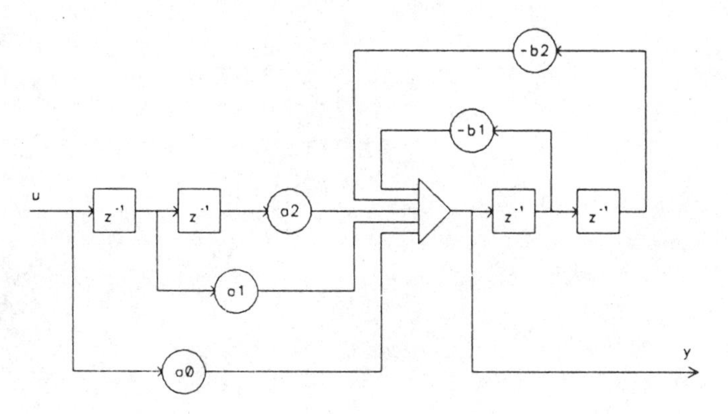

Figure 4.1 Second-order digital filter in direct form

(2) However filters are more often realised in practice in the so-called canonical form, which is arrived at by rewriting eqn (4.1) as two eqns:

$$\frac{v(z)}{u(z)} = \frac{1}{1 + b_1z^{-1} + b_2z^{-2}} \tag{4.2}$$

$$\frac{y(z)}{v(z)} = a_0 + a_1z^{-1} + a_2z^{-2} \tag{4.3}$$

Clearly the product of eqns (4.2) and (4.3) gives eqn (4.1) but the two components may be taken separately, as seen in figure 4.2(a), (b), and combined to give figure 4.2(c). This is an alternative to figure 4.1 that is commonly preferred.

(3) The third option to be discussed [Ag75, Go85, Go89a] is obtained from $F(z)$ by substituting $(\delta + 1)$ for z and arranging the result in powers of δ^{-1} to give

$$F(\delta) = \frac{c_0 + c_1\delta^{-1} + c_2\delta^{-2}}{1 + r_1\delta^{-1} + r_2\delta^{-2}} \tag{4.4}$$

The relationship between c_i, r_i and a_i, b_i depends upon the order of the filter; in this case

$$c_0 = a_0$$

$$c_1 = 2a_0 + a_1 \qquad r_1 = 2 + b_1$$

$$c_2 = a_0 + a_1 + a_2 \qquad r_2 = 1 + b_1 + b_2$$

The filter described by eqn (4.4) is then realised canonically as in figure 4.3 to reach its final form.

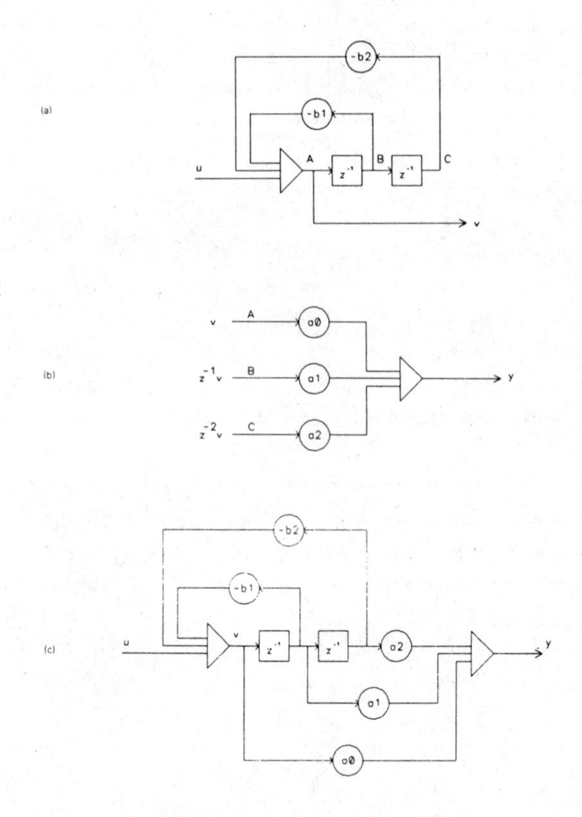

Figure 4.2 Second-order digital filter in canonical form

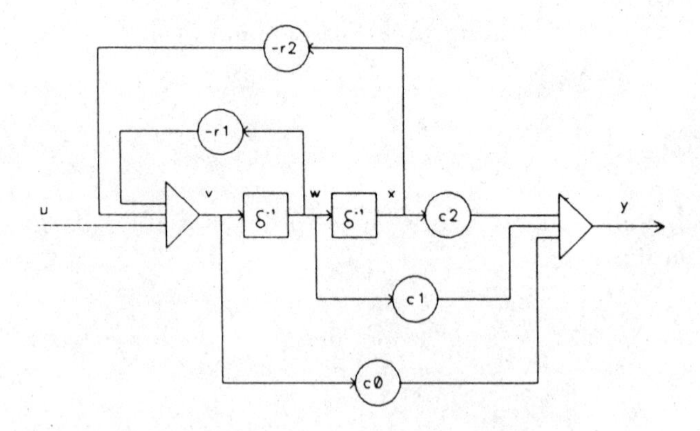

Figure 4.3 Second-order digital filter in delta form

The reason for the substitution of $(\delta + 1)$ for z may be understood by writing

$$\delta = z - 1$$

$$\therefore \quad \delta^{-1} = \frac{1}{z-1} = \frac{z^{-1}}{1-z^{-1}} = \frac{u_o(z)}{u_i(z)}$$

Figure 4.4 shows δ^{-1} to be an accumulator; the quantity to be entered into memory at the beginning of each sampling interval is the sum of the latest sample and the previous summation. In software terms, δ^{-1} requires FETCH, ADD and STORE, rather than the FETCH and STORE of z^{-1}, so there is a slight time-penalty implied in its use. This however is more than compensated for, in most circumstances, by other features of its performance which are discussed in detail in Chapter 5.

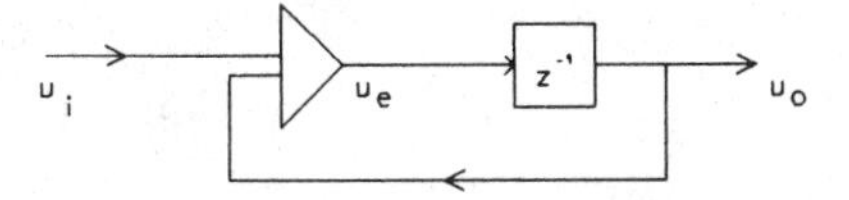

Figure 4.4 The operator δ^{-1} expressed in terms of z^{-1} to show that it is an accumulator

Although one might realise a given transfer function in a dozen structures and only three are being considered here, there is little loss of generality. Other structures that are not considered directly will be found to exhibit the salient characteristics of one (or more) of those described.

4.3 Variation of filter error with sampling interval

One of the most fundamental decisions to be made is on the sampling rate. It is often one of the most difficult, since frequently people do not know what factors should determine their decision. In many cases, perhaps in most, the choice is made 'blind' by deciding on a sampling interval equal to about one-tenth of the smallest system time constant, or perhaps a sample frequency of ten times the system bandwidth. A rule-of-thumb often works well enough to put you in the right region, but would you benefit by increasing or decreasing T? The issue is often settled in practice by nothing more elegant than trial-and-error. To improve on that, one must give informed consideration to the relevant factors: computer wordlength, number of bits in the ADC, filter structure, the plant to be controlled. In the sections that follow these things are examined to discover how they affect the choice of sampling interval and the errors that are generated.

All the error types enumerated are affected by choice of sampling interval to some degree, but in different ways; as sampling frequency is increased, some errors are reduced, some become larger, while others hardly change. The filter-structure may also enter the picture, with an increase in T producing greater error with one structure but less with another.

To clarify a complex state of affairs we consider each error type in turn and ask how a change in T affects it, for each of the three filter structures.

4.3.1 e_a: algorithmic error

This does not depend upon the structure and always increases with T. Intuition alone would lead one to that conclusion, but it is possible to quantify the error.

If the analogue filter and its digital equivalent are given by

$$F(s) = \frac{n_0 + n_1 s + n_2 s^2}{m_0 + m_1 s + m_2 s^2}$$

and

$$F(z) = \frac{a_0 + a_1 z^{-1} + a_2 z^{-2}}{1 + b_1 z^{-1} + b_2 z^{-2}}$$

then the error coefficients J_i (section 4.1.1) may be calculated in the following way, as explained in [Fo86].

Deciding arbitrarily to compute the first five terms of e_a (that is, J_0 to J_4), define the 5 × 5 matrices:

$$N = \begin{bmatrix} n_0 & & & & \\ n_1 & n_0 & & & \\ n_2 & n_1 & n_0 & & \\ 0 & n_2 & n_1 & n_0 & \\ 0 & 0 & n_2 & n_1 & n_0 \end{bmatrix}$$

$$M = \begin{bmatrix} m_0 & & & & \\ m_1 & m_0 & & & \\ m_2 & m_1 & m_0 & & \\ 0 & m_2 & m_1 & m_0 & \\ 0 & 0 & m_2 & m_1 & m_0 \end{bmatrix}$$

$$\mathcal{T} = \begin{bmatrix} 1 & & & & \\ & T & & & \\ & & T^2/2! & & \\ & & & T^3/3! & \\ & & & & T^4/4! \end{bmatrix}$$

$\mathscr{L}$ has dimensions 5×3 (because five J-coefficients are requested and there are three a-coefficients) and is defined:

$$\mathscr{L} = \begin{bmatrix} 1 & 1 & 1 \\ 0 & -1 & -2 \\ 0 & 1 & 4 \\ 0 & -1 & -8 \\ 0 & 1 & 16 \end{bmatrix}$$

Succeeding columns of $\mathscr{L}$ (when required) consist of powers of -3, -4, etc. It can be shown that

$$\begin{aligned} J &= [J_0\ J_1\ J_2\ J_3\ J_4]^T \\ &= \mathscr{L}\, B - \mathscr{T}^{-1}\, N^{-1}\, M\, \mathscr{T}\, \mathscr{L}\, A \end{aligned} \qquad (4.5)$$

$$\text{where } A = [a_0\ a_1\ a_2]^T$$
$$B = [1\ \ b_1\ b_2]^T.$$

The extension of the matrices to any dimensions is obvious.

Although eqn (4.5) may look a little daunting, it is in fact very easy to program, and if one of the mathematical software packages is available (MATLAB or MATHCAD, for example) then evaluation of J becomes trivial. If you are working without such assistance, note that it is not necessary to perform any matrix inversions. The elements of $\mathscr{T}^{-1}$ are merely the inverse of the elements of $\mathscr{T}$, while N^{-1} may be found by a simple process that takes advantage of the form assumed by N, as explained in [Fo86]. Computing the values of the J_i allows us to evaluate the series

$$e_a = J_0 u + J_1(T) \overset{(1)}{u} + J_2(T^2/2!) \overset{(2)}{u} + \ldots$$

to any given number of terms, if we make some sensible assumptions about u. This is discussed at greater length in section 4.6.1.

4.3.2 e_s: sampler quantisation error

Like e_a, this is independent of filter structure, but *unlike* e_a it has minimal dependence upon T.

This error too can be quantified; the method is given below without attempting to explain the underlying theory, which can be found in [Fo89], but it will be noted that again the computations required are easily carried out.

Consider an arbitrary filter

$$F(z) = \frac{a_0 + a_1 z^{-1} + \ldots + a_n z^{-n}}{1 + b_1 z^{-1} + \ldots + b_n z^{-n}}$$

and for simplicity let us suppose the poles to be distinct. In that case:

$$F(z) = r_0 + \frac{r_1}{1 - p_1 z^{-1}} + \frac{r_2}{1 - p_2 z^{-1}} + \ldots + \frac{r_n}{1 - p_n z^{-1}} \tag{4.6}$$

If the sampler rounds off, rather than truncates, the quantisation effect can be represented as zero mean random noise lying in the range ±0.5 lsb, all values equally probable. Given that e_i and e_s represent the rms input and output noise respectively, we may write

$$e_\mathrm{s}^2 = \Phi_0 \, e_\mathrm{i}^2 \tag{4.7}$$

where Φ_0 is determined by the filter transfer function. Representing 1 lsb by Δ, the probability density function $p(x)$ for the input noise x is shown in figure 4.5.

Since it is certain that the value of the input noise lies between $\pm\Delta/2$, the area under $p(x)$ equals unity and the height of the rectangle in figure 4.5 must be $1/\Delta$. Thus we have the well known result

$$\begin{aligned} e_\mathrm{i}^2 &= \int_{-\Delta/2}^{\Delta/2} x^2 p(x)\,\mathrm{d}x \\ &= \frac{1}{\Delta}\int_{-\Delta/2}^{\Delta/2} x^2\,\mathrm{d}x \\ &= \Delta^2/12 \end{aligned} \tag{4.8}$$

Substituting eqn (4.8) into (4.7):

$$e_\mathrm{s} = \Delta[\Phi_0/12]^{1/2} \tag{4.9}$$

It has been shown [Fo89] that Φ_0 can be evaluated in terms of the parameters r_i and p_i in eqn (4.6) as

$$\Phi_0 = [r_0\ r_1 \ldots r_n]\,[\mathrm{Q}]\,[r_0\ r_1 \ldots r_n]^T \tag{4.10}$$

where

$$Q = \begin{bmatrix} 1 & 1 & 1 & 1 & \cdot \\ 1 & \dfrac{1}{1 - p_1^2} & \dfrac{1}{1 - p_1 p_2} & \dfrac{1}{1 - p_1 p_3} & \cdot \\ 1 & \dfrac{1}{1 - p_2 p_1} & \dfrac{1}{1 - p_2^2} & \dfrac{1}{1 - p_2 p_3} & \cdot \\ \cdot & \cdot & \cdot & \cdot & \cdot \end{bmatrix}$$

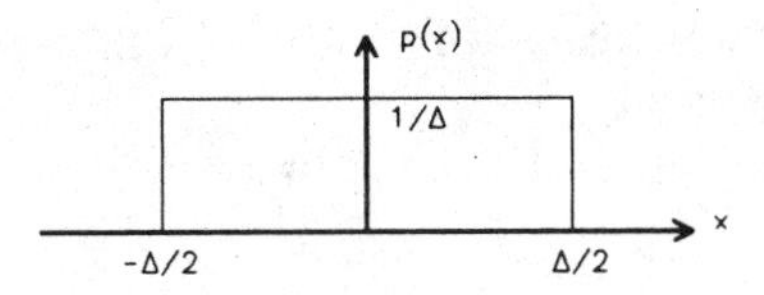

Figure 4.5 The probability density function $p(x)$ for the random variable x

This result, as pointed out above, is easily programmed so the error due to quantisation of the ADC is readily determined for a given filter from eqns (4.9) and (4.10).

4.3.3 e_r: multiple word truncation error

This type of error is dependent upon structure.

For the canonical and delta forms, e_r behaves likes e_s but is of much lesser magnitude when handled correctly; it can therefore be disregarded when the filter is realised properly in these structures. Filter realisation to minimise e_r is not a simple matter and details of how it may be accomplished will be examined in Chapter 5.

For the direct form, however, the case is different. As T is reduced, e_r *increases* and may become a major source of error. For that reason, this structure is best avoided. It has been used throughout the previous three chapters because it seems the obvious, or natural, implementation of a transfer function; unfortunately it has the drawback noted. The error analysis that follows makes these points.

Consider the three structures in turn.

(a) Direct form

For simplicity and clarity, ideas are developed initially around a first-order filter, the subsequent extension to the nth-order being trivial.

With infinite precision available:

$$y(k) = a_0u(k) + a_1u(k-1) - b_1y(k-1)$$

where

$u(k)$ is the current sample of the filter input,
$y(k)$ is the current sample of the filter output.

In an alternative notation that is found to be convenient because it is brief:

$$y_0 = a_0u_0 + a_1u_1 - b_1y_1 \tag{4.11}$$

where the subscript 0 indicates current value, the subscript 1, the first or most recent past value, etc.

With finite precision eqn (4.11) becomes

$$(y_0 + \gamma_0) = (a_0 + \alpha_0)(u_0 + \mu_0) + (a_1 + \alpha_1)(u_1 + \mu_1) - (b_1 + \beta_1)(y_1 + \gamma_1) + \tau_0 \quad (4.12)$$

The quantities γ, α, μ, β are the quantisation errors incurred in representing y, a, u and b, respectively, while τ_0 is the error generated when the multilength products are reduced to a lesser length. To illustrate this point, suppose that the filter coefficients and input data were each held to a single wordlength (which is often done, though it will be shown later that this is unlikely to be the best course of action); then the products $(a_0 + \alpha_0)(u_0 + \mu_0)$ and $(a_1 + \alpha_1)(u_1 + \mu_1)$ would be stored in full in double-length words. If the summation $(y_0 + \gamma_0)$ is, say, triple length, the product $(b_1 + \beta_1)(y_1 + \gamma_1)$ will be quadruple length, like all terms $(b_n + \beta_n)(y_n + \gamma_n)$ in the general case. The summation which constitutes the right-hand side of eqn (4.12), excluding τ_0, is properly therefore of quadruple length and must be reduced to triple, an operation which is accomplished by the addition of τ_0 with a value equal and opposite to that of the fourth word of the summation. τ_0 may therefore in this example have any value between ± 1 lsb of the triple-length word. Note that if the reduction is undertaken *before* the summation, then τ_0 will be larger, ranging over $\pm$n lsb where n is the number of quantities summed.

A lot more is said about the internal arrangements for the arithmetic in Chapter 5, but the broad description just given is sufficient for the argument that follows.

Multiplying out and rearranging the terms of eqn (4.12) leads to

$$(y_0 + (b_1 + \beta_1)y_1) + (\gamma_0 + (b_1 + \beta_1)\gamma_1) = (a_0u_0 + a_1u_1) + (\alpha_0u_0 + \alpha_1u_1) + (a_0\mu_0 + a_1\mu_1) + (\alpha_0\mu_0 + \alpha_1\mu_1) + \tau_0 \quad (4.13)$$

Re-stating eqn (4.13) in terms of the z-transform:

$$y(z)(1 + (b_1 + \beta_1)z^{-1}) + \gamma(z)(1 + (b_1 + \beta_1)z^{-1}) = u(z)((a_0 + \alpha_0) + (a_1 + \alpha_1)z^{-1}) + \mu(z)((a_0 + \alpha_0) + (a_1 + \alpha_1)z^{-1}) + \tau_0(z) \quad (4.14)$$

At this point let us generalise and define

$$A(z) = a_0 + a_1z^{-1} + \ldots$$

$$\alpha(z) = \alpha_0 + \alpha_1z^{-1} + \ldots$$

$$A^+(z) = A(z) + \alpha(z)$$

$$B(z) = 1 + b_1z^{-1} + \ldots$$

$$\beta(z) = 0 + \beta_1z^{-1} + \ldots$$

$$B^+(z) = B(z) + \beta(z)$$

From eqn (4.14) therefore, dropping the (z) for convenience:

$$y + \gamma = u\left[\frac{A^+}{B^+}\right] + \mu\left[\frac{A^+}{B^+}\right] + \tau\left[\frac{1}{B^+}\right]$$

But since

$$y = u\left[\frac{A}{B}\right] \tag{4.15}$$

therefore

$$\gamma = u\left[\frac{A^+}{B^+} - \frac{A}{B}\right] + \mu\left[\frac{A^+}{B^+}\right] + \tau\left[\frac{1}{B^+}\right] \tag{4.16}$$

(b) Canonical form

The filter in this structure is described by two equations which, in the first-order case, would be

$$v_0 = u_0 - b_1v_1$$

$$y_0 = a_0v_0 + a_1v_1$$

Introducing quantisation error in the manner described above, leads to:

$$(v_0 + (b_1 + \beta_1)v_1) + (\zeta_0 + (b_1 + \beta_1)\zeta_1) = u_0 + \mu_0 + \tau_0$$

$$\begin{aligned}(y_0 + \gamma_0) &= (a_0 + \alpha_0)(v_0 + \zeta_0) + (a_1 + \alpha_1)(v_1 + \zeta_1)\\ &= (a_0v_0 + a_1v_1) + (a_0\zeta_0 + a_1\zeta_1) + (\alpha_0v_0 + \alpha_1v_1)\\ &\quad + (\alpha_0\zeta_0 + \alpha_1\zeta_1)\end{aligned}$$

where ζ is the quantisation error on v. Hence

$$\gamma_0 = (a_0 + \alpha_0)\zeta_0 + (a_1 + \alpha_1)\zeta_1 + \alpha_0v_0 + \alpha_1v_1$$

or in terms of the operator z:

$$\gamma(z) = A^+(z)\zeta(z) + \alpha(z)v(z) \tag{4.17}$$

The previous analysis, culminating in eqn (4.16), is relevant here if $A = 1$ and $\alpha = 0$. Thus

$$\zeta(z) = u(z)\left[-\frac{\beta(z)}{B(z)B^+(z)}\right] + \mu(z)\left[\frac{1}{B^+(z)}\right] + \tau\left[\frac{1}{B^+(z)}\right] \tag{4.18}$$

Since

$$v(z) = u(z)/B(z) \tag{4.19}$$

one can substitute from eqns (4.18) and (4.19) into (4.17) to produce

$$\begin{aligned}\gamma &= u\left[-\frac{A^+\beta}{BB^+}\right] + \mu\left[\frac{A^+}{B^+}\right] + \tau\left[\frac{A^+}{B^+}\right] + u\left[\frac{\alpha}{B}\right] \\ &= u\left[\frac{B\alpha - A\beta}{BB^+}\right] + \mu\left[\frac{A^+}{B^+}\right] + \tau\left[\frac{A^+}{B^+}\right] \\ &= u\left[\frac{A^+}{B^+} - \frac{A}{B}\right] + \mu\left[\frac{A^+}{B^+}\right] + \tau\left[\frac{A^+}{B^+}\right]\end{aligned} \tag{4.20}$$

This is different from eqn (4.16) only in the final term.

(c) Delta form

Recall that the basic element of the filter in this formulation is not the shift-register but the accumulator. It follows that, although the analysis below is carried out in much the same way as that for the canonical form, the filter and its performance are very different. Using a first-order example as before, the two equations necessary to describe the filter operation are

$$\begin{aligned} v_0 &= u_0 - r_1 w_0 \\ y_0 &= c_0 v_0 + c_1 w_0 \end{aligned} \tag{4.21}$$

If the quantisation errors on v, c, r and w are ζ, σ, ρ and η respectively, then eqns (4.21) lead to

$$\begin{aligned} &(v_0 + (r_1 + \rho_1)w_0) + (\zeta_0 + (r_1 + \rho_1)\eta_0) = u_0 + \mu_0 + \tau_0 \\ &(y_0 + \gamma_0) = (c_0 v_0 + c_1 w_0) + (c_0 \zeta_0 + c_1 \eta_0) \\ &+ (\sigma_0 v_0 + \sigma_1 w_0) + (\sigma_0 \zeta_0 + \sigma_1 \eta_0) \end{aligned} \tag{4.22}$$

Also

$$w = \Sigma\, v_0 \qquad \text{or} \qquad w(z) = \delta^{-1} v(z)$$

$$\eta = \Sigma\, \zeta_0 \qquad \text{or} \qquad \eta(z) = \delta^{-1} \zeta(z)$$

so that, adopting operator notation and generalising:

$$v(z) + \zeta(z) = (u(z) + \mu(z) + \tau(z))/R^{+}(\delta)$$

where

$$R^{+}(\delta) = 1 + (r_1 + \rho_1)\delta^{-1} + (r_2 + \rho_2)\delta^{-2} + \ldots$$

and

$$\begin{aligned} y(z) + \gamma(z) &= (C(\delta) + \sigma(\delta))(v(z) + \zeta(z)) \\ &= (C(\delta) + \sigma(\delta))(u(z) + \mu(z) + \tau(z))/R^{+}(\delta) \end{aligned}$$

Since

$$y(z) = u(z)C(\delta)/R(\delta)$$

therefore, dropping the arguments as before:

$$\gamma = u\left[\frac{C^{+}}{R^{+}} - \frac{C}{R}\right] + \mu\left[\frac{C^{+}}{R^{+}}\right] + \tau\left[\frac{C^{+}}{R^{+}}\right] \tag{4.23}$$

where

$$R = 1 + r_1\delta^{-1} + r_2\delta^{-2} + \ldots$$

$$C = c_0 + c_1\delta^{-1} + c_2\delta^{-2} + \ldots$$

$$\sigma = \sigma_0 + \sigma_1\delta^{-1} + \sigma_2\delta^{-2} + \ldots$$

$$\rho = 0 + \rho_1\delta^{-1} + \rho_2\delta^{-2} + \ldots$$

The global quantisation error created in the chosen filter structures is set out in eqns (4.16), (4.20) and (4.23), and could be determined in a similar manner for the many alternatives. The quantity τ, the error created by truncating a multiple-length product to a lesser length (sometimes referred to as 'round-off error' [Ja72]), will only be appreciable in the canonical and delta forms if the implementation is incorrectly undertaken. Provided therefore the appropriate measures are taken to minimise τ, it will be negligible in comparison to μ, for these two structures. In the case of the direct form, however, the amplification of τ may be considerable (see eqn (4.16)), leading to the conclusion that it is unwise to use this structure, certainly when the processor wordlength is small. Consider, for example, the low pass filter used earlier:

$$F(s) = \frac{1}{1 + s}$$

Using the bilinear transform, with $T = 0.05$ sec, this becomes

$$F(z) = \frac{0.02439(1 + z^{-1})}{1 - 0.95122z^{-1}}$$

In the steady state ($z = 1$), $B = 0.04878$, and τ is therefore amplified by a factor of about 20.

Thus the quantisation error created by a digital filter may, with the reservation noted above, be attributed chiefly to the relatively coarse quantisation of the ADC (converters longer than 12 bits are not common and are expensive) and to filter coefficient approximation or representational error; on that basis, eqns (4.20) and (4.23) can be written without the third term:

$$\gamma_c = u\left[\frac{A^+}{B^+} - \frac{A}{B}\right] + \mu\left[\frac{A^+}{B^+}\right] = \gamma_{cc} + \gamma_{cn} \tag{4.24}$$

for the canonical form, and

$$\gamma_d = u\left[\frac{C^+}{R^+} - \frac{C}{R}\right] + \mu\left[\frac{C^+}{R^+}\right] = \gamma_{dc} + \gamma_{dn} \tag{4.25}$$

for the delta form. The first term in each case gives the error due to coefficient approximation, while the effect of ADC quantisation is described by the second.

4.3.4 e_c: coefficient representational error

This form of error is also dependent upon structure.

For the direct and canonical forms, e_c *increases* with sampling frequency; the greater the order of the filter, the more pronounced the effect.

The delta structure is, in contrast, very much less sensitive to coefficient representation, allowing operation to much higher sampling frequencies than the other two structures.

High order filters involve high powers of T, so their coefficients will begin to disappear rapidly as T becomes small. This effect is counteracted by realising a high order filter as a number of low order filters in series or parallel (corresponding to the expression of the filter transfer function as a product of factors, or in the form of partial fractions). It is accepted that in direct or canonical forms, the factors or fractions may only be of the first or second order, but third-order sections have been realised in delta form [Go85] very satisfactorily. Fourth-order sections would prove practicable in many cases.

For the direct and canonical forms the relevant expression, from eqn (4.24), is

$$e_c = u\left[\frac{A^+}{B^+} - \frac{A}{B}\right] = u[E_{cc}] \tag{4.26}$$

For the delta form, from eqn (4.25):

$$e_c = u\left[\frac{C^+}{R^+} - \frac{C}{R}\right] = u[E_{dc}] \tag{4.27}$$

This expression is only what common sense would lead one to expect; the most significant thing about it, in a control context, is likely to be the low-frequency or steady-state gain. It was demonstrated in section 4.1.2 that an error of about 8% was created by 8-bit representation of the coefficients in

$$F(z) = \frac{0.024390(1 + z^{-1})}{1 - 0.951220z^{-1}} \tag{4.28}$$

which is the bilinear transformation of

$$F(s) = \frac{1}{s + 1}$$

The delta form of the same filter is found from eqn (4.28), by substituting $z = \delta + 1$, and is

$$F(\delta) = \frac{0.024390 + 0.048780\delta^{-1}}{1 + 0.048780\delta^{-1}}$$

Representing the coefficients once more to 8 bits we have

$$F_8(\delta) = \frac{0.0234375 + 0.046875\delta^{-1}}{1 + 0.046875\delta^{-1}} \tag{4.29}$$

The zero frequency gain in the delta form is found by setting $\delta = 0$, and it can be seen from eqn (4.29) that this leads to a value of unity. There is *no* zero frequency error. This will apply to any filter with a zero frequency gain of 1, however gross the coefficient approximations. When the gain is required to be something other than 1, the filter is best represented as a unity gain section followed by a constant multiplier. In this way, the steady-state error will always be very small – no greater than the representational error in the constant multiplier.

From eqn (4.26)

$$E_{cc} = \frac{A^+}{B^+} - \frac{A}{B}$$

$$= \frac{(A + \alpha)B - (B + \beta)A}{BB^+}$$

$$= \frac{B\alpha - A\beta}{BB^+}$$

$$= \frac{1}{B^+}\left(\alpha - \beta\frac{A}{B}\right) \tag{4.30}$$

In a similar manner we derive from eqn (4.27):

$$E_{dc} = \frac{C^+}{R^+} - \frac{C}{R}$$

$$= \frac{1}{R^+}\left(\sigma - \rho\frac{C}{R}\right) \tag{4.31}$$

The values of σ, ρ, α and β will change as T is changed, but in a manner that shows no dependence upon T; it is merely random. The only variation that *is* dependent upon T is that of B^+. It is not obvious how R^+ could be expected to change, but in fact it does not behave like B^+ – it remains constant as T declines.

The result is that E_{cc} shows a random variation superimposed upon a curve that is rising with increasing sampling frequency, while E_{dc} displays a similar random quality but no sensible variation with sampling rate.

The choice of a specific value for coefficient wordlength is considered in detail in Chapter 5 because it impinges on overflow and underflow problems in the internal arithmetic, that is, on the value of e_r. To reduce this error (e_r) to negligible proportions in some given set of circumstances, it will be found that the coefficients must be expressed to a certain accuracy; thus the curve for e_c will be determined fairly directly by the problem specification if the design procedure described in Chapter 5 is carried through.

4.3.5 Summary

The material of section 4.3 is summarised in figure 4.6, which indicates what happens to the various types of error as the sampling rate is increased.

4.4 Sampling interval and loop stability

Some pains were taken in the previous section to explain the way in which the sampling interval affects the errors generated within the digital filter. There is however a further important consideration.

Error type	Structure		
	Direct	Canonical	Delta
e_c	increases	increases	no increase
e_r	increases	—	—
e_s	little change	independent of structure	
e_a	falls sharply		

Figure 4.6 Summary of error variation with increase in sampling frequency

The mere existence of the ZOH within the controlling loop, the generation of those flat-topped pulses, has a de-stabilising effect that must be recognised. When the sampling rate is high, the staircase waveform is comprised of very small steps and the loss of stability is negligible, but it may be the case that we wish to sample as slowly as is consistent with adequate control. How slowly might that be?

4.4.1 Gauging the reduction in stability

The longer the sampling interval, the greater will be the reduction in system stability, and the more severe the problem of compensation that must be tackled by the controller. The plant, by virtue of its inherent characteristics, presents a certain difficulty in control; to this is added what is in effect a time delay due to the action of the ZOH. To sample rapidly, and effectively remove the latter, is one course of action often favoured, but there may be substantial savings to be made by accepting a slower sampling rate with a less powerful processor, always provided that the exacerbated control problem can be handled.

An effective way of examining the degradation in stability for a given sampling interval has already been introduced, though not for that purpose. The technique used to carry out design in the w-plane is all we need. The plant $G(s)$ is converted to $G_z(z)$, as described in Chapter 3, and then returned to the w-plane using the bilinear transform (or whatever method of emulation is favoured) to give $G_z(w)$. The frequency response of $G_z(w)$ may then be plotted, on a Nichols chart for example, to give the gain and phase margins. These are a simple and effective guide to relative stability; any change that reduces these margins is de-stabilising and the magnitude of the reduction is a direct measure of the amount of de-stabilisation.

It will be found that, for small values of T, the difference between the frequency responses of $G(s)$ and $G_z(w)$ is negligible. As T is increased, the difference grows and the margins are reduced from their original, analogue, values, to become at some point intolerably small. It is a matter of judgement to decide when T has reached its largest permissible value, but typically one would not allow the margins to shrink by more than about 30%.

4.5 Rationale for the choice of system parameters

It is clear that there is a complex interaction between the choices that the designer has to make: the processor wordlength, the number of bits in the A/D converter, the algorithm chosen for the controller, the structure in which it is realised, and of course the value of T.

How might one choose a path through this labyrinth? A rational approach is to choose T from considerations of loop stability, as explained in section 4.4, and to form the other choices in consequence of this. The total error e_t is the sum of e_a, e_s and e_c; assuming that we avoid the use of the direct form and properly organise the internal arithmetic, e_r will be negligible. As the sampling frequency is changed the three errors that dominate are found to vary, typically in the manner of figure 4.7. This shows three sets of curves for each of e_s, and e_c, and the curve for e_a, in the case of a compensator designed to improve the dynamics of the pitch

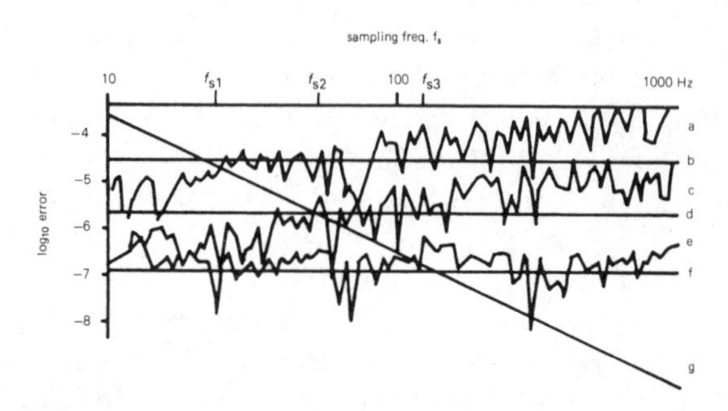

Figure 4.7 The errors e_c, e_s and e_a graphed against sampling frequency for various combinations of ADC, processor wordlength, and filter structure. The graphs are identified as follows: (a) e_c, with 12-bit processor, floating-point arithmetic, canonical form; (b) e_s, for 8-bit ADC; (c) e_c, with 16-bit processor, floating-point arithmetic, canonical form; (d) e_s, for 12-bit ADC; (e) e_c, with 16-bit processor, floating-point arithmetic, delta form; (f) e_s, for 16-bit ADC; (g) e_a, the algorithmic error

control loop of a light aircraft. The curves of e_s correspond to 8-, 12- or 16-bit A/D converters, and those of e_c to 12- and 16-bit computations with the canonical form, and a 16-bit computation with the delta form. (The determination of these curves will be dealt with in section 4.6.)

The value of T or f_s, selected on grounds of loop stability, identifies a frequency on the f_s-axis; a vertical line through this point intersects the error curves at certain levels, prompting certain conclusions. For example, if the sampling frequency were f_{s1} then an 8-bit ADC with a 12-bit processor would match the algorithmic error very well, as e_a, e_s and e_c would all contribute about equally to the level of error generated.

At a higher sampling rate f_{s2}, the error will be reduced by an order of magnitude if a 12-bit ADC is used, but only if the processor is changed to 16 bits.

The error may be reduced by a further order of magnitude if the ADC works to 16 bits by sampling at f_{s3}, provided the compensator algorithm is structured in delta form; the 16-bit processor is still sufficient however. (The A320 Airbus uses a 16-bit ADC/processor system while another 'digital' aircraft of earlier vintage uses a 12-bit system.)

All three of these outline solutions have the virtue of being consistent, but the 'best' choice can only be made with full knowledge of the circumstances obtaining in a particular application. If, for example, the error levels associated with the first and least expensive solution were acceptable, then the cost of the third option would be unjustified.

As well as highlighting consistent choices, figure 4.7 brings out inconsistencies. If, for example, one were sampling at f_{s1}, it would not make sense to use a 16-bit processor or a 12-bit ADC.

4.6 Numerical evaluation of e_a, e_c, e_s

In order to graph the plots illustrated in figure 4.7, one must at certain points make some assumptions. These are discussed.

4.6.1 e_a*, the algorithmic error*

This is defined in section 4.1.1 as

$$e_a = J_0 u + J_1(T) \overset{(1)}{u} + J_2(T^2/2!) \overset{(2)}{u} + \ldots$$

where u is the input to the filter, $\overset{(n)}{u}$ is the nth derivative of u, and the coefficients J_i are defined by eqn (4.5).

A means has been provided in section 4.3 for computing the values of the coefficients, but what figures are to be used for the derivatives?

A reasonable course of action is to assume that the input to the filter, the digital controller, is a sinusoid having the natural frequency of the system. The amplitude of the sinewave must be chosen carefully to be neither too large nor too small, bearing in mind the circumstances of the application. In this way one may evaluate the terms in e_a and hence the rms value attained by e_a in the course of its cyclic variation. A computer program may be written to plot this quantity against T, and this is the curve e_a in figure 4.7. The natural frequency in this example is about 6 rad/sec and a suitable value for the magnitude of the sinewave is considered, from knowledge of the aircraft, to be around 1.5°.

4.6.2 e_c, *the coefficient representational error*

The error in representing a given coefficient to some stated number of bits is quite random and might be anything from zero to 1 lsb, with equal probability; changing the value of T merely creates a new random error. However it is declared in eqns (4.26) and (4.30) that for the canonical structure, the error in the filter output on account of the coefficient approximation is given by

$$uE_{cc} = \frac{u}{B^+}\left(\alpha - \beta\frac{A}{B}\right)$$

For the delta structure, eqns (4.27) and (4.31) state that

$$uE_{dc} = \frac{u}{R^+}\left(\sigma - \rho\frac{C}{R}\right)$$

u is the filter input, the sinewave described in the previous section, and the other symbols all represent polynomials in z^{-1}. For the chosen frequency of input sinewave, the magnitude or rms value of the sinusoidal output error may therefore be calculated.

4.6.3 e_s, *sampler quantisation error*

In section 4.3.2 it is explained that e_s is the rms noise output from the filter due to the quantisation level Δ chosen for the ADC. e_s is given, to repeat eqns (4.9) and (4.10), by the expressions

$$e_s = \Delta[\Phi_0/12]^{1/2}$$

$$\Phi_0 = [r_0\ r_1\ \ldots\ r_n]\ [\mathrm{Q}]\ [r_0\ r_1\ \ldots\ r_n]^T$$

$$Q = \begin{bmatrix} 1 & 1 & 1 & 1 & \cdot \\ 1 & \dfrac{1}{1 - p_1^2} & \dfrac{1}{1 - p_1 p_2} & \dfrac{1}{1 - p_1 p_3} & \cdot \\ 1 & \dfrac{1}{1 - p_2 p_1} & \dfrac{1}{1 - p_2^2} & \dfrac{1}{1 - p_2 p_3} & \cdot \\ \cdot & \cdot & \cdot & \cdot & \cdot \end{bmatrix}$$

The filter has poles $p_1, p_2, \ldots p_n$ with residues $r_1, r_2, \ldots, r_n$. When the numerator and denominator of the filter have the same order, then r_0 will exist (see eqn (4.6)), but if it does not then the first row and column of Q will be absent.

For a given value of T, the poles and residues of the filter can be calculated and e_s evaluated over a range of T, for the relevant values of Δ. Thus the curves e_s are plotted. The transfer function employed in the example of figure 4.7 was

$$C(s) = 0.65 \left[\frac{1 + 0.167s}{1 + 0.40s} \right]$$

transferred to the z-domain using the bilinear transform.

4.7 Conclusions

The output of a digital algorithm will always be different from that of its analogue counterpart (however you choose to define that). This difference is the error in the algorithm output, and in this chapter we have examined the factors that cause it; in particular we have focussed on the way in which the errors change with the choice of sampling interval and have seen that there will always be some optimum value for the sampling frequency, given a particular hardware configuration. Alternatively, the same process of analysis can be used to guide the choice of hardware elements that go to make up the system; possibly the latter function is the more important.

The subject has not proved easy to investigate (and probably not to read); much remains still to be understood but a picture can be discerned. We hope that its essential elements have been conveyed to the reader, both to explain the role of the sampling interval in determining the performance of a digital filter and to guide the choice of sampling interval at the design stage.

5 Requirements for Implementation

This chapter and Chapter 6 are structured with the intention of leading the reader through the essential processes on the way to successful and effective implementation. In this chapter we show how control strategies which have been synthesised by the methods of the preceding chapters may now be translated into general functional and computational requirements. This is a vital process which is ideally carried out prior to deciding upon a specific processor around which to design a digital controller. Chapter 6 then deals with the more specific issues of hardware and software. Throughout both chapters, a strongly design-orientated approach is maintained in order that the reader not only learns practical design methods, but also gains an appreciation of the important issues concerned with the implementation of digital controllers.

This chapter firstly elaborates upon the comparison between analogue and digital forms of implementation, because it is important that an appreciation of the reasons for using digital control is not lost. In order that we can determine the essential questions which need answering, a generalised *architectural* scheme of a processor system is next introduced (in advance of Chapter 6, which deals more specifically with the hardware). After that, we present a generalised *functional* scheme with a view to identifying the overall requirements such as input/output, control structure, compensation stages, etc. From the basis provided by these generalised considerations we move on to a detailed discussion of filter structures presented earlier, in Chapter 4. This enables us to convert the discrete transfer functions into equations and to determine the precision required for the computations.

5.1 Analogue *vs* digital controllers

Some of the issues relating to the 'analogue or digital?' question were raised in Chapter 1. What we intend here is an elaboration of those issues in order to reflect further upon the motivation for using a digital controller, and to concentrate the designer's mind upon the important aspects of implementation.

Analogue controllers are relatively simple with low component costs and cheap manufacturing technology, although as the number of functions increases so does the complexity. Changes to control parameters can be made by altering component values, and within certain constraints, different control strategies can be achieved by altering the types of component, for example, changing a resistor to a capacitor can give integral control.

Even the most simple digital controller must have certain essential features in addition to the processor itself (see the next section), and the basic components are considerably more expensive than their analogue counterparts. The manufacturing process is significantly more complicated, with multilayer printed circuit boards containing multiple track layouts for data and address busses. Since the functions are carried out sequentially by the processor under the control of the software, the hardware complexity of the controller does not necessarily increase significantly as the number of functions increases, although the extra computation must still be carried out within the sample period. Changes to control parameters (and even to the overall control strategy) can be made by altering the software, although this may not necessarily be as straightforward as it sounds. A Microprocessor Development System is an essential tool for this to be carried out efficiently.

It is possible to make a qualitative comparison between analogue and digital controllers, and table 5.1 gives most of the important advantages

Table 5.1 Comparison between analogue and digital controllers

	Advantages	*Disadvantages*
Analogue	Infinite resolution	Drift
	No time delays	Non-repeatability
	Simple PCB design	Adaptive control difficult
	Cheap components	Interlocking needs separate logic circuit
	Easy commissioning	Dedicated to a specific application
	Low development costs	Single ic solutions difficult
Digital	Programmability	Sample delays
	Adaptive control possible	Computation delays
	Interlocking logic easy	Complex PCB design
	Adaptable to various applications	Commissioning needs planning
	Single ic solutions possible	High development overheads

and disadvantages which are often quoted. Some however, while being strictly true, are not really applicable, and it is worth discussing these in a control system context in order that the reader may gain a balanced view.

A good example is the often stated disadvantage of thermal drift in analogue circuits, and the corresponding lack of it in digital solutions. In practice, the problem relates almost exclusively to the signal conditioning which amplifies the transducer outputs to appropriate levels for the control circuit, a function which is required by both analogue and digital controllers. Once conditioned high-level signals (for example, ± 10 V) have been generated, thermal drift in an analogue control circuit is not a problem. Even analogue integrators with hold times in excess of an hour can readily be designed with modern operational amplifiers.

A doubtful advantage frequently attributed to a digital solution is that of programmability, making it easily adjustable to a variety of applications. Either the program memory must be reprogrammed, or the change can be made from a terminal via a serial communications link, in which case there will normally be an extra hardware and software overhead.

In any case, it is really quite simple and quick to make changes to an analogue circuit. Clearly the situation would be different for more profound changes in control strategy rather than simple parameter changes, but a digital solution may still require extra data memory, program memory and computational capacity, and possibly extra analogue input/output.

The most likely reasons for using a digital controller were given in Chapter 1:

- lower cost (usually only for complex controllers)
- the need for adaptive control
- the need for a single integrated circuit solution in high-volume applications.

It is widely recognised that implementing the numerical processes involved in digital controllers is not straightforward, and many try to avoid the problems by using a high-level language with floating point arithmetic. However it is inevitably necessary to 'buy in' extra processing power to achieve this, so the approach may be inconsistent with the aim of reducing cost. (In any case, the use of floating-point arithmetic does not automatically overcome all the problems.) If lower cost is the motivation, then computational overkill is inappropriate. If sophisticated control strategies are wanted, then it is desirable to maximise the computation time available for identification, adaptation and other advanced concepts in control by improving computational efficiency. If the aim is for a single chip solution, then it will be necessary to optimise the functionality in order to minimise the size of the silicon. Consequently, whatever the application or motivation for using a digital controller, we recommend that the computational requirements are carefully assessed. This will ensure that computation is minimised either by guiding the choice of variable types in a high-level language, or by specifying the precision of low-level assembler routines.

Getting digital controllers to work properly under all circumstances is not straightforward, so it is vital to understand the underlying processes which are operating within the controller. The following sections tackle these issues and give a reliable methodology for accurate and efficient implementation.

5.2 Generalised architecture of a digital controller

Figure 5.1 shows the essential components of a digital controller. To some extent what is illustrated is a fairly normal microprocessor system, and assumes a standard architecture with address, data and control busses, although other architectures having instruction busses and pipelining capabilities are also possible; the general scheme however does not depend upon the specifics of the architecture. The important thing to appreciate is how the requirements for implementing digital control relate to the generality of digital processor systems, and conversely how the various parts of the overall scheme fit within a control perspective. It is also necessary to perceive the complete scheme in order to identify important questions which need to be answered before a digital controller can effectively be designed.

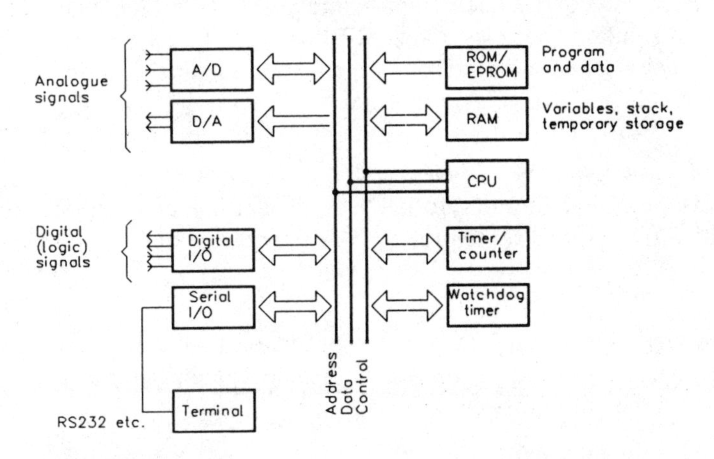

Figure 5.1 Generalised digital controller

The scheme itself is divided into two parts as shown in the figure – the internal functions (including the processor itself) to the right, and the peripherals to the left. Explanations of these, with a definition of their functions, are as follows:

CPU (Central Processing Unit)
The core of the controller which carries out the arithmetic and logical manipulation, controls access to and from the RAM and timer(s), and handles data transfer to and from the peripherals (all under the influence of a program usually stored in ROM/EPROM).

RAM (Random Access Memory)
The basic store for variables, intermediate results, etc. Also used for the stack which stores return addresses during sub-routine and interrupt calls.

Timer/counter(s)
Used for timing internal processes and/or external events; especially useful in the present context to set the sampling rate for the control algorithms; can usually be configured to act as counters on external inputs.

ROM/EPROM (Read Only Memory/Erasable Programmable ROM)
Stores the control program itself, as well as data for arithmetic constants, look-up tables, etc.

A/D (Analogue-to-Digital converter(s))
Converts analogue feedback signals into equivalent digital words of a suitable resolution (such as 8 bits, 12 bits, etc.). May be a number of separate converters, a single converter with a separate multiplexer or a multichannel device having an internal multiplexer.

D/A (Digital-to-Analogue converter(s))
Converts the digital output(s) from the controller into analogue signals. As with the A/D, various configurations are possible.

Digital I/O
Provides logic signals for interlocking: inputs for monitoring limit switches and control signals, outputs for such things as enabling power amplifiers, switching indicator lights, etc.

Serial I/O (an optional feature)
Gives the facility to make changes to parameters if required by means of a serial communications link to an external terminal (or a computer with a mass storage device).

Watchdog timer (an optional feature)
Provides a means of protection against 'glitches' which cause mis-operation of the software. If the timer is not regularly reset by the CPU, it will reset the whole controller so that it re-starts the software.

Consideration of the basic scheme identifies a number of important questions that need to be answered:

1. What type of processor is needed? For example, data word size, address space, and instruction set must be considered.
2. How much memory is needed? The size of RAM for variables and PROM for program must be determined.
3. What about the analogue input/output requirements? How many channels are needed? What is the resolution of the conversion process, and the speed of operation? What basic sample rates are needed, and how much computation needs to be done within the sample period?
4. What about the development environment? – that is, what facilities are needed? how should the software be structured? what languages are to be used? etc.

Some of these questions have already been considered in earlier chapters; others will be addressed in the ensuing sections.

5.3 Functional representation

It is possible to think about the digital controller from a generalised functional point of view (figure 5.2). This really represents the starting point for implementation and can be seen to consist of a number of inputs, a number of discrete transfer functions (possibly cascaded or paralleled) and a number of outputs derived from the discrete transfer functions. It is obvious that within this generalised functional scheme there will be input variables, output variables and also intermediate variables between the internal functions. These variables may need to be combined by means of addition or subtraction, and sometimes may need to be multiplied by a constant gain factor. There will also be temporary results from calculating the responses of the discrete transfer functions. Less obvious is the fact that some of these results must be stored as internal variables (this is explained in section 5.4). In addition, there will be coefficients by which the variables are multiplied within the discrete transfer functions. In a practical application the variables, the intermediate results and the coefficients are all quantised.

It is useful to reflect upon the types of function to be implemented, which can be grouped into three categories:

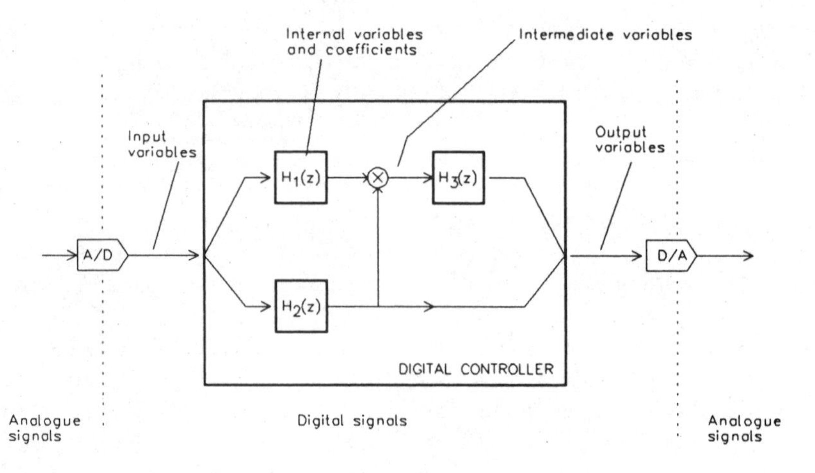

Figure 5.2 Generalised functional scheme

1. *Analogue input and output.* These are 'low-level' types of function which will be hardware-dependent. Although the devices are reviewed in Chapter 6, together with the position of their associated software procedures within the overall software structure, anything more detailed is more appropriately dealt with in other texts [Co87].
2. *Generation of control system error(s).* This is a straightforward function, possibly multiprecision if the input/output wordlength exceeds that of the processor being used, but nevertheless needing nothing unusual in the way of computation.
3. *Control loop compensators.* In some cases the compensator may be a simple gain, in other cases a discrete transfer function, possibly also including a gain. A simple gain must inherently be greater than unity, such that overflow is inevitable for errors exceeding the proportional band; this potential for overflow must be detected and correctly limited, but otherwise a simple gain is merely a multiplication. However, the implementation of a discrete transfer function is a much more complex affair, and must be well understood and correctly designed if it is to work properly.

The emphasis will be upon the issues concerned with the compensator implementation – if these are correctly tackled, then the other computational functions should not pose any extra problems.

From this starting point it is possible to define what we need in the way of key performance requirements, which can be listed as follows:

1. A specified accuracy for the overall performance of the control strategy.
2. An adequate response to large inputs, in other words minimising the effects of saturation and overflow. Note that there may well be some

controller saturation inherent in the size of the control gains (see section 5.7.2), but we must ensure that the digital implementation does not create additional problems.

3. An adequate response to small inputs, in other words ensuring that a minimal change at the input must create an appropriate change at the output. (It is not always recognised that this must be considered with digital controllers, but lack of response to small inputs is analogous to effects such as backlash, which are known to be highly detrimental to closed loop systems.)

Certain decisions, which are clearly related to these requirements, will already have been taken on the basis of 'system' considerations and were discussed in Chapter 4. These are:

(a) *The wordlength for the input and output variables*. An assessment is made of the quantisation effects on the input and output in terms of accuracy and noise. This is used to specify the resolution of A/D and D/A conversion. It clearly makes sense that the intermediate variables within the controller should have a wordlength consistent with that of the input/output variables.

(b) *The sample rate*. This is determined (see Chapter 4) to meet certain overall system requirements of performance and/or stability. Its value is, of course, used in determining the coefficients of the discrete transfer functions which are to be implemented. In some cases there may be a number of different sample rates associated with different feedback loops.

Additional considerations which must be taken account of during implementation mainly relate to the internal processes of calculating the discrete transfer functions, and can be summarised as follows:

(a) *The accuracy of the coefficients*. Determination of this relates mainly to the first of the three requirements, that is, to achieving a specified performance accuracy.

(b) *The wordlength for the internal variables*. This will be based upon that of the inputs and outputs, but in the general case we need to allow for the possibility of both overflow and underflow. These two factors determine the overall size of the internal variables, and it will emerge that they relate strongly to the requirements 2 and 3 listed above, although obviously they will also have a general effect upon the overall performance accuracy.

What we therefore need to do is to determine the precision to which the computations within the controller must be carried out. This will give us the minimum required wordlengths, and can be used to ensure that suitable numerical routines are used.

5.4 Filter structures

The published literature on the subject of recursive digital filters includes a large number of possible structural forms. These structures reflect the ways in which the discrete transfer functions can be interpreted both diagrammatically and in terms of the equations to be implemented. Since we are concentrating here on the basics, just two structures will be considered at this point – the direct form and canonic form. An important alternative, the delta form, is considered in depth in section 5.9.

The following discussion of filter structures is applied to first- and second-order discrete transfer functions; higher order transfer functions can, of course, be achieved by cascaded combinations.

Compensators that include an integral term are a special case for a number of reasons, and are dealt with separately in section 5.8.

5.4.1 The direct form

A general second-order discrete transfer function may be written as

$$\frac{y(z)}{x(z)} = \frac{a_0 + a_1 z^{-1} + a_2 z^{-2}}{1 + b_1 z^{-1} + b_2 z^{-2}}$$

The direct form is the most obvious or direct way of implementing a digital filter (and hence the name), and is found by simply cross-multiplying the terms in the discrete transfer function equation to give

$$y(z)(1 + b_1 z^{-1} + b_2 z^{-2}) = x(z)(a_0 + a_1 z^{-1} + a_2 z^{-2})$$

which transforms to the time domain expression:

$$y(k) + b_1 y(k-1) + b_2 y(k-2) = a_0 x(k) + a_1 x(k-1) + a_2 x(k-2)$$

or using the notation introduced in section 4.3.3:

$$y_0 + b_1 y_1 + b_2 y_2 = a_0 x_0 + a_1 x_1 + a_2 x_2$$

The unknown in the equation is of course y_0, and re-arranging to make this variable the subject of the equation gives

$$y_0 = a_0 x_0 + a_1 x_1 + a_2 x_2 - b_1 y_1 - b_2 y_2$$

Figure 5.3 is a diagrammatic representation of this equation. When it comes to programming the filter, certain essential equations are implicit in the digrammatic representation. The set of equations which needs to be implemented at each sample instant is shown in the fragment of C program given below. Note that this is incomplete, because of the lack of variable declarations and because three procedures are not defined (their purposes should be clear from their names); the objective is simply to demonstrate the sequence of calculations.

```
main()
{
      /*variable declarations omitted*/
      while (time < end_of_time)
      {
            if new_sample()
            {
                  x0= input_adc();
                  y0= a0*x0 + a1*x1 + a2*x2 - b1*y1 - b2*y2;
                  output_dac(y0);
                  /* shift the variables ready for next sample */
                  x2 = x1;
                  x1 = x0;
                  y2 = y1;
                  y1 = y0;
                  time = time + T;
            }
      }
}
```

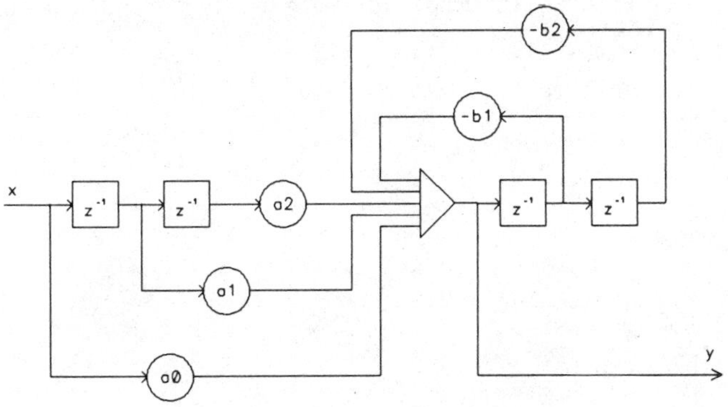

Figure 5.3 Direct-form *z*-filter

5.4.2 *Canonic form*

This is found by introducing a new variable v such that

$$\frac{v(z)}{x(z)} = \frac{1}{1 + b_1 z^{-1} + b_2 z^{-2}}$$

Cross-multiplying as before, and making v the subject of the equation gives

$$v = x - b_1 z^{-1} v - b_2 z^{-2} v$$

which transforms to

$$v_0 = x_0 - b_1 v_1 - b_2 v_2$$

The output equation is then given by

$$y_0 = a_0 v_0 + a_1 v_1 + a_2 v_2$$

Figure 5.4 is a diagrammatic representation of the canonic form. Again the important thing is the set of equations to be implemented, which is once more identified using a C program:

```
main()
{
     /*variable declarations omitted */
     while (time < end_of_time)
     {
            if new_sample()
            {
                 x0 = input_adc();
                 v0 = x0 - b1*v1 - b2*v2;
                 y0 = a0*v0 + a1*v1 + a2*v2;
                 output_dac(y0);
                 /*shift the internal variable ready for next sample */
                 v2 = v1;
                 v1 = v0;
                 time = time+T;
            }
      }
}
```

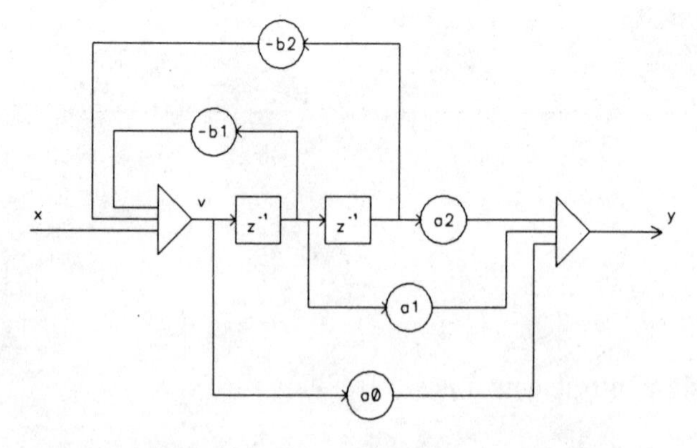

Figure 5.4 Canonic-form z-filter

Either the direct or the canonic form of the digital filter may be used, but it is widely recognised that the canonic form has certain advantages. Although two equations are needed for the canonic form instead of one for the direct form, there are fewer stored variables and shift operations. Also the effects of truncation in the direct form have been shown in Chapter 4 to be more significant. For this reason, subsequent consideration will be restricted to the canonic form, although the principles can be applied equally well to other forms.

It should be stressed at this stage that, if there were a ready supply of cheap, fast, high-resolution floating-point processors which could be efficiently programmed in a high-level language, then the equations written above would be all that is necessary. Even though processor technology continues to advance, in practice such devices are not available, and it is impossible to ignore quantisation and other effects in digital filters.

5.5 Determining the precision of computation

It is sometimes stated that the choice of processor determines the wordlengths for the internal computations, and analytical or simulation methods have been developed for predicting the effects of quantisation, demonstrated by frequency responses, time responses or by the movement of the *z*-domain roots (see, for example, [Li71]). Such an approach, however, has the wrong emphasis for the purposes of actually *designing* a controller (as opposed to *analysing* an existing design).

The question that needs to be answered is not 'If we quantise this coefficient or that variable to some given precision, how will it affect the performance?' but rather 'If we want performance to this accuracy, to what precision must we carry out the internal computations?'. Clearly, the analytical methods referred to above can be *indirectly* used to answer the second question, but the design rules which we describe may be used *directly* to specify the precision. This will influence the choice of processor from the outset (Chapter 6 elaborates on this).

The design approach which we present for specifying the computational precision is complementary to the better known analytical methods, and not in any way contradictory – indeed, it can be used to infer their results. The principles are simple to understand, and the procedures themselves are algebraic in nature. They yield useful quantitative values without over-constraining the design, and give consistency with the stated aim of maintaining a design-orientated approach.

The following two sections firstly address the determination of coefficient wordlength, and then move on to the internal variable wordlength. Together they define fully the computational requirements, although the way their conclusions are translated for programming is left until the software considerations are dealt with in Chapter 6.

5.6 Coefficient wordlength

The compensators in the controller are described by discrete transfer functions, derived either by emulation of an equivalent continuous transfer function or by formulation in the *w*- or *z*-domains. Whichever method of design has been used, what we really need to know is how inaccuracies caused by quantisation in the coefficients of the discrete transfer function affect the overall performance of the closed loop control system. However, while this can be predicted in a specific application, it is impossible in a generalised sense, and so we are restricting ourselves to the problem of achieving a specified performance accuracy in the controller, leaving it up to the designer to relate this to the system as a whole.

We must therefore decide how to quantify the controller's performance, and we offer a general-purpose method which relates the accuracy of the discrete filter coefficients to the accuracy of the coefficients of an equivalent continuous filter. We mentioned above that there are other better known methods based upon time or frequency responses, but pointed out that, while these are useful for analysis, they are less useful for synthesis. It is worth remarking that if one knows the way in which the equivalent continuous filter's coefficients are affected, one can always deduce the effect in the time or frequency domain. The general-purpose nature of our method is further emphasised by the fact that some of the coefficients are themselves likely to be meaningful characteristics such as d.c. gain, high-frequency gain, time constants, resonant frequency, etc., all of which relate directly to time and/or frequency responses.

The idea of an equivalent continuous transfer function of course raises the question of the transform method to be used. However, at high sample rates for which the coefficient sensitivity problem has been shown to be most acute, the various methods of transforming between the *z*- and *s*-domains give nearly the same results. It is therefore sensible to choose a transform method which is easy to use, and the algebraic nature of the bilinear transform makes it a natural choice, although the sensitivity analysis which follows can readily be re-worked for an alternative transform if desired. Often the specification for the accuracy of a controller's parameters will in any case not be particularly precise, and will usually have greater uncertainties than those introduced by the difference between transform methods. At low sample rates, the method is still applicable in practice even with a greater divergence between transform methods – this is because the 'sensitivity factors' which are then derived are much lower, and hence accuracy is not as important.

The fact that the method requires the discrete compensator to have an equivalent continuous filter does not restrict its use to designs formulated by emulation from the *s*-domain. Any discrete compensator can readily be transformed into the *s*-domain; what matters is its performance in the time or frequency domain, either of which is readily derived from the *s*-domain.

5.6.1 *General procedure for choice of coefficient wordlength*

The main stages in the process we describe are as follows:

1. Determine algebraic expressions for the equivalent continuous filter coefficients in terms of the discrete coefficients and the sampling interval using the bilinear transformation.
2. Calculate the sensitivity of the equivalent continuous filter coefficients to changes in the discrete coefficients, by means of a matrix of sensitivity factors as explained below.
3. Use the sensitivity factors to determine the permitted variations in the discrete coefficients in order to achieve a specified accuracy in the continuous coefficients.
4. Convert the permitted variations into equivalent wordlengths, and calculate approximated values for the discrete coefficients.

5.6.2 *Sensitivity analysis of a first-order filter*

The general form of corresponding continuous and discrete first order transfer functions can be written as:

$$H(s) = \frac{n_0 + n_1 s}{1 + m_1 s} \qquad H(z) = \frac{a_0 + a_1 z^{-1}}{1 + b_1 z^{-1}}$$

Using the bilinear transform $s = 2(z - 1)/T(z + 1)$, it is possible to inter-relate the coefficients of the two transfer functions. Table 5.2 lists expressions which give both the discrete coefficients as functions of the continuous coefficients, and *vice versa*. The former are used to calculate *exact values* for the discrete coefficients when using an emulation technique; the latter will be used to deduce sensitivity factors by which the *accuracy* of the approximated discrete coefficients can be determined.

Table 5.2 Emulation table for first-order z-filters (canonic)

$H(s) = \dfrac{n_0 + n_1 s}{1 + m_1 s}$	$H(z) = \dfrac{a_0 + a_1 z^{-1}}{1 + b_1 z^{-1}}$
$a_0 = \dfrac{n_0 T + 2n_1}{T + 2m_1}$	$n_0 = \dfrac{a_0 + a_1}{1 + b_1}$
$a_1 = \dfrac{n_0 T - 2n_1}{T + 2m_1}$	$n_1 = \left(\dfrac{a_0 - a_1}{1 + b_1}\right)\dfrac{I}{2}$
$b_1 = \dfrac{T - 2m_1}{T + 2m_1}$	$m_1 = \left(\dfrac{1 - b_1}{1 + b_1}\right)\dfrac{I}{2}$

It is possible to define the coefficient sensitivity by means of a matrix S of sensitivity factors:

$$\begin{bmatrix} \dfrac{\delta n_0}{n_0} \\[2ex] \dfrac{\delta n_1}{n_1} \\[2ex] \dfrac{\delta m_1}{m_1} \end{bmatrix} = [S] \begin{bmatrix} \dfrac{\delta a_0}{a_0} \\[2ex] \dfrac{\delta a_1}{a_1} \\[2ex] \dfrac{\delta b_1}{b_1} \end{bmatrix}$$

in which

$$[S] = \begin{bmatrix} s_{n_0,a_0} & s_{n_0,a_1} & s_{n_0,b_1} \\ s_{n_1,a_0} & s_{n_1,a_1} & s_{n_1,b_1} \\ s_{m_1,a_0} & s_{m_1,a_1} & s_{m_1,b_1} \end{bmatrix}$$

Notice that the matrix S relates small fractional changes in the discrete coefficients to fractional changes in the continuous coefficients. The sensitivity factors within S are therefore normalised, and can be interpreted as follows:

Sensitivity factor = 1 ('normal' sensitivity)
Sensitivity factor » 1 ('high' sensitivity)
Sensitivity factor « 1 ('low' sensitivity)

Taking the sensitivity for coefficient n_0 first, a standard mathematical expression for small changes in n_0 is

$$\delta n_0 = \frac{\partial n_0}{\partial a_0}\delta a_0 + \frac{\partial n_0}{\partial a_1}\delta a_1 + \frac{\partial n_0}{\partial b_1}\delta b_1$$

which can be re-written in terms of fractional changes as

$$\frac{\delta n_0}{n_0} = \left(\frac{\partial n_0}{\partial a_0}\frac{a_0}{n_0}\right)\frac{\delta a_0}{a_0} + \left(\frac{\partial n_0}{\partial a_1}\frac{a_1}{n_0}\right)\frac{\delta a_1}{a_1} + \left(\frac{\partial n_0}{\partial b_1}\frac{b_1}{n_0}\right)\frac{\delta b_1}{b_1}$$

and hence the sensitivity factors can be deduced. For example:

$$s_{n_0,a_0} = \frac{\partial n_0}{\partial a_0}\cdot\frac{a_0}{n_0}$$

$$s_{n_0,a_1} = \frac{\partial n_0}{\partial a_1}\cdot\frac{a_1}{n_0} \qquad \text{etc.}$$

From table 5.2:

$$n_0 = \frac{a_0 + a_1}{1 + b_1}$$

from which

$$\frac{\partial n_0}{\partial a_0} = \frac{1}{1 + b_1}$$

and hence

$$s_{n_0, a_0} = \frac{1}{1 + b_1} \cdot \frac{a_0}{n_0} = \frac{a_0}{a_0 + a_1}$$

This can be evaluated directly, but it is also useful to express it in terms of the equivalent continuous filter parameters:

$$s_{n_0,\, a_0} = \frac{1}{1 + \dfrac{T - 2m_1}{T + 2m_1}} \cdot \frac{\dfrac{n_0 T + 2n_1}{T + 2m_1}}{n_0} = \frac{n_0 T + 2n_1}{2Tn_0}$$

The virtue of an algebraic expression is that the trend as the sampling period decreases can be seen – in this case it is approximately inversely proportional to T, a trend which is not obvious from the expression written in terms of the discrete coefficients.

Notice that it is equally possible, as explained below, to express the sensitivity in terms of other characteristics of the equivalent continuous filter which more specifically relate to its purpose, rather than in terms of its basic coefficients. This is particularly relevant when the discrete compensator design is based on an emulation approach. The following example illustrates this, and also demonstrates how the sensitivity factors will be used in practice.

Example 5.1: A phase advance compensator

This can be expressed by the transfer function

$$H(s) = G\left[\frac{1 + sk\tau}{1 + s\tau}\right]$$

$$= \frac{n_0 + n_1 s}{1 + m_1 s}$$

$$= n_0\left[\frac{1 + sn_1/n_0}{1 + sm_1}\right]$$

Thus G = low-frequency gain = n_0
k = phase advance ratio = n_1/n_0m_1
τ = time constant = m_1.

Using the procedure described above, the following sensitivity matrix can be derived:

$$\begin{bmatrix} \dfrac{\delta G}{G} \\ \dfrac{\delta k}{k} \\ \dfrac{\delta \tau}{\tau} \end{bmatrix} = \begin{bmatrix} \dfrac{T + 2k\tau}{2T} & \dfrac{T - 2k\tau}{2T} & \dfrac{2\tau - T}{2T} \\ \dfrac{T^2 - 4k^2\tau^2}{4kT\tau} & \dfrac{4k^2\tau^2 - T^2}{4kT\tau} & \dfrac{T^2 - 4\tau^2}{4T\tau} \\ 0 & 0 & \dfrac{4\tau^2 - T^2}{4T\tau} \end{bmatrix} \begin{bmatrix} \dfrac{\delta a_0}{a_0} \\ \dfrac{\delta a_1}{a_1} \\ \dfrac{\delta b_1}{b_1} \end{bmatrix}$$

(the derivation of this is given in Appendix F).

Note that the sensitivity matrix can be expressed equally well in terms of the discrete coefficients, in which case the sensitivity factors become independent of the sample period T, a surprising result perhaps. However, expressing the matrix as shown gives the factors in terms of the more immediate parameters of the continuous filter.

Taking specific values, $G = 1$, $k = 5$ and $\tau = 0.1$ sec with a sample period T of 0.01 sec (that is, 100 Hz sampling), gives the following numerical values for the exact coefficients and for their sensitivities:

$$\left.\begin{aligned} a_0 &= 4.8095 \\ a_1 &= -4.7143 \\ b_1 &= -0.9048 \end{aligned}\right\} \text{from table 5.2}$$

$$\begin{bmatrix} \dfrac{\delta G}{G} \\ \dfrac{\delta k}{k} \\ \dfrac{\delta \tau}{\tau} \end{bmatrix} = \begin{bmatrix} 50.5 & -49.5 & 9.5 \\ -49.995 & 49.995 & -9.975 \\ 0 & 0 & 9.975 \end{bmatrix} \begin{bmatrix} \dfrac{\delta a_0}{a_0} \\ \dfrac{\delta a_1}{a_1} \\ \dfrac{\delta b_1}{b_1} \end{bmatrix}$$

These values have been derived from the algebraic expressions given above. They can also be derived numerically by making small changes to the discrete coefficients and re-calculating the equivalent continuous coefficients; expressing the resulting change in fractional terms gives the corresponding sensitivity factor. This alternative method can readily be programmed into a computer, although observation of 'trends' by inspecting the algebraic expressions is lost.

The reader should note that the sensitivity of the low-frequency gain G is dependent not only upon a_0 and a_1, but also quite strongly upon b_1. And yet the expression for calculating the exact value for b_1, derived from table 5.2, is

$$b_1 = \frac{T - 2\tau}{T + 2\tau}$$

which would suggest that G and b_1 are independent. The expression for G, derived from the *reverse* transformation, makes the dependency quite clear, however:

$$G = \frac{a_0 + a_1}{1 + b_1}$$

This kind of observation (which frequently applies in discrete filter implementations) stresses the importance of deriving the sensitivity factors in the manner shown (that is, using the continuous filter coefficients expressed as functions of the discrete coefficients).

A brief reflection on the purpose of sensitivity analysis is worthwhile here. It is reasonably obvious that the maximum error in a coefficient will be half the size of its least significant bit, and so these factors could be used to answer the question: 'If x, y, z are the coefficient wordlengths, what would be the implementation accuracy?' However from a design point of view this is not much use, and in any case this question could have been answered without recourse to the sensitivity analysis. As we have observed before, the important question is posed the other way round: 'If such and such an implementation accuracy is required, what must the coefficient word length be?' – that is, synthesis rather than analysis.

Given that the final error will be a combination of the errors from the individual discrete coefficients (taking into account their respective sensitivity factors), there are a large number of 'degrees-of-freedom' in the design process due to the possibility of unequal contributions to the total error. It is also necessary to decide whether to use a 'worst-case' or a statistical approach to the problem (as was done at one point in Chapter 4, for example). For simplicity we suggest a straightforward approach which has neither the pessimism of the worst-case analysis nor the complexity of a full statistical analysis, and experience has shown that this will quickly produce good working figures for the wordlength. A more rigorous approach is open to readers who wish to pursue it.

Returning to the example, supposing the compensator needs to be implemented to a 5% accuracy in its parameters G, k, τ (typically what would be specified in a control application, although differing individual accuracies for the three parameters could be specified if necessary). We will work on the basis of each discrete coefficient having the same fractional or percentage accuracy, which implies the same number of *significant* bits in the coefficient. (An alternative would have been the same absolute accuracy, implying the same number of *fractional* bits, that is, bits following the binary point). Dividing the required accuracy for each continuous coefficient by the sum of the moduli of its sensitivity factors would give a worst-case value for the fractional accuracy, but this is often unduly pessimistic. A simple but sensible alternative is to divide the required accuracy by the root of the sum of the squares (rss) of the coefficients' sensitivity factors, which makes some concession to the statistical nature of the problem without getting too complicated. Using this approach gives the following results:

5% accuracy in G requires 0.07% coefficient accuracy, that is

$$\frac{5}{[50.5^2 + (-49.5)^2 + 9.5^2]^{1/2}} \%$$

5% accuracy in k requires 0.07% coefficient accuracy

5% accuracy in τ requires 0.5% coefficient accuracy.

We should choose the greatest accuracy requirement of 0.07% although it is worth observing that the lower accuracy for τ only involves b_1; since the sensitivity of G and k to changes in b_1 is much lower than for the other two discrete coefficients, it would probably be possible to work with this lower accuracy for b_1 if desired. However keeping the same accuracy for all the coefficients leads to the following specifications for the coefficient values and their accuracies:

$$\begin{aligned} a_0 &= 4.8095 \pm 0.0034 \\ a_1 &= -4.7143 \pm 0.0033 \\ b_1 &= -0.9048 \pm 0.00063 \end{aligned}$$

The coefficients with their accuracies can be translated into corresponding numbers of bits using the inequalities given below:

For a_0: $0.0034 \geq 1/2^{n+1}$, that is, 8 fractional bits
$4.8095 \leq 2^m$ 3 integer bits

For a_1: $0.0033 \geq 1/2^{n+1}$, that is, 8 fractional bits
$4.7143 \leq 2^m$ 3 integer bits

For b_1: $0.00063 \geq 1/2^{n+1}$, that is, 10 fractional bits
$0.908 \leq 2^m$ no integer bits

Now that the resolution has been determined, the nearest values of the coefficients can be calculated. This calculation is relatively obvious, but is included below:

<nearest_value> = (integer (<exact_value> × 2^n + 0.5))/2^n

that is, $a_0 = 4.80859375\ (100.11001111_B)$

$a_1 = -4.71484375\ (100.10110111_B)$

$b_1 = -0.9052734375\ (0.1110011111_B)$

To complete the process, it is useful to calculate corresponding values for the continuous filter parameters and their errors. These are:

$G = 1.072$ (7.2% error)

$k = 4.996$ (0.1% error)

$\tau = 0.1017$ (1.7% error)

This example has yielded one result which by chance exceeded the specified 5% accuracy. Ultimately it is up to the designer what to do about this – either leave it or add an extra fractional bit. Conversely, there might be a tendency to try and take advantage of the small error in k by selectively reducing the wordlength of one of the coefficients; however an unforeseen adjustment to parameters in the compensator during testing of the control system might easily reverse the errors, giving G a very small error and k a much larger error, so this tendency should be resisted. The procedure described specifies wordlengths which should satisfy all combinations of compensator parameters, except of course where the sample period is fundamentally altered.

5.6.3 Sensitivity analysis of a second-order filter

Generalised forms for second-order continuous and discrete filters can be written as

$$H(s) = \frac{n_0 + n_1 s + n_2 s^2}{1 + m_1 s + m_2 s^2} \qquad H(z) = \frac{a_0 + a_1 z^{-1} + a_2 z^{-2}}{1 + b_1 z^{-1} + b_2 z^{-2}}$$

Exactly the same procedure as before can be used to derive a sensitivity matrix. The algebra is more complicated, but table 5.3 gives expressions for the coefficients as with the first-order filter, and this greatly facilitates the process. A full sensitivity matrix would be 5 × 5, but an example is probably a better way to illustrate the analysis and the way the results will be used.

Table 5.3 Emulation table for second-order z-filters

$H(s) = \dfrac{n_0 + n_1 s + n_2 s^2}{1 + m_1 s + m_2 s^2}$	$H(z) = \dfrac{a_0 + a_1 z^{-1} + a_2 z^{-2}}{1 + b_1 z^{-1} + b_2 z^{-2}}$
$a_0 = \dfrac{n_0 T^2 + 2n_1 T + 4n_2}{T^2 + 2m_1 T + 4m_2}$	$n_0 = \dfrac{a_0 + a_1 + a_2}{1 + b_1 + b_2}$
$a_1 = \dfrac{2n_0 T^2 - 8n_2}{T^2 + 2m_1 T + 4m_2}$	$n_1 = \dfrac{a_0 - a_2}{1 + b_1 + b_2} T$
$a_2 = \dfrac{n_0 T^2 - 2n_1 T + 4n_2}{T^2 + 2m_1 T + 4m_2}$	$n_2 = \dfrac{a_0 - a_1 + a_2}{1 + b_1 + b_2} \dfrac{T^2}{4}$
$b_1 = \dfrac{2T^2 - 8m_2}{T^2 + 2m_1 T + 4m_2}$	$m_1 = \dfrac{1 - b_2}{1 + b_1 + b_2} T$
$b_2 = \dfrac{T^2 - 2m_1 T + 4m_2}{T^2 + 2m_1 T + 4m_2}$	$m_2 = \dfrac{1 - b_1 + b_2}{1 + b_1 + b_2} \dfrac{T^2}{4}$

Example 5.2: A notch filter, emulated design

Notch filters are quite commonly used in electro-mechanical control systems in order to eliminate the excitation of a structural resonance. They will generally have low damping factors (typically in the region of 0.05) and the required accuracy for the cut-off frequency ω_0 may be relatively high. A standard expression for a second-order notch filter is:

$$H(s) = \frac{1 + s^2/\omega_0^2}{1 + 2\zeta s/\omega_0 + s^2/\omega_0^2}$$

$$n_0 = 1 \qquad m_1 = 2\zeta/\omega_0$$

$$n_1 = 0 \qquad m_2 = n_2$$

$$n_2 = 1/\omega_0^2$$

This can be translated into a discrete transfer function using the emulation table, and gives the following expressions for the discrete transfer function and its coefficients:

$$H(z) = \frac{a_0 + a_1 z^{-1} + a_0 z^{-2}}{1 + a_1 z^{-1} + b_2 z^{-2}}$$

$$a_0 = \frac{T^2 + 4n_2}{T^2 + 2m_1T + 4n_2} \qquad a_1 = \frac{2T^2 - 8n_2}{T^2 + 2m_1T + 4n_2}$$

$$b_2 = \frac{T^2 - 2m_1T + 4n_2}{T^2 + 2m_1T + 4n_2}$$

A feature of this particular filter requirement, which was not evident in the first-order example, is that the values of two pairs of the discrete transfer function coefficients are repeated (a_0 and a_2 are the same, and a_1 and b_1 are the same). This has been recognised in the above expressions by using a_0 twice and a_1 twice in the discrete transfer function. This will help with sensitivity because it reduces the number of independent coefficient variations which are possible. (It also helps with the algebra because it reduces the number of terms to be calculated!)

Notice that, whereas in the first-order example the analysis was presented in terms of the more meaningful parameters of the phase advance, this has not been done here, mainly because it is less tractable algebraically because of the square root relationship between n_2 and cut-off frequency. It is however reasonably easy to infer the sensitivity of parameters such as cut-off frequency from the sensitivity factors which will be derived.

The converse expressions giving the continuous filter coefficients, taking advantage of the 'pairing' effect, are as follows:

$$n_0 = \frac{2a_0 + a_1}{1 + a_1 + b_2} \qquad n_2 = \frac{2a_0 - a_1}{1 + a_1 + b_2} \cdot \frac{T^2}{4}$$

$$n_1 = 0 \qquad m_1 = \frac{1 - b_2}{1 + a_1 + b_2} \cdot T$$

A sensitivity matrix can now be derived:

$$\begin{bmatrix} \dfrac{\delta n_0}{n_0} \\[2ex] \dfrac{\delta n_2}{n_2} \\[2ex] \dfrac{\delta m_1}{m_1} \end{bmatrix} = \begin{bmatrix} \dfrac{T^2 + 4n_2}{2T^2 n_0} & 0 & -\dfrac{(T^2 - 2m_1T + 4n_2)}{4T^2 n_0} \\[2ex] \dfrac{T^2 + 4n_2}{8n_2} & \dfrac{16n_2^2 - T^4}{8T^2 n_2} & -\dfrac{(T^2 - 2m_1T + 4n_2)}{4T^2} \\[2ex] 0 & \dfrac{4n_2 - T^2}{2T^2} & -\dfrac{(T + m_1)(T^2 - 2m_1T + 4n_2)}{4T^2 m_1} \end{bmatrix} \begin{bmatrix} \dfrac{\delta a_0}{a_0} \\[2ex] \dfrac{\delta a_1}{a_1} \\[2ex] \dfrac{\delta b_2}{b_2} \end{bmatrix}$$

Inspection of the sensitivity factors shows that most are now approximately inversely proportional to the *square* of the sample period T, which indicates that in general the sensitivity of the second-order filter is likely to be significantly worse than the first-order example previously examined – the numerical values will confirm this.

Now let us consider a specific filter having a 1 Hz cut-off frequency with a damping factor of 0.05. This can readily be shown to have a bandwidth between −3 dB points of 0.1 Hz (that is, 0.05 Hz either side of the centre frequency). Again using $T = 0.01$ sec:

$$H(s) = \frac{1 + 0.0253s^2}{1 + 0.0159s + 0.0253s^2}$$

Substituting into the algebraic matrix above gives numeric values as follows:

$$\begin{bmatrix} \frac{\delta n_0}{n_0} \\ \frac{\delta n_2}{n_2} \\ \frac{\delta m_1}{m_1} \end{bmatrix} = \begin{bmatrix} 506.5 & 0 & -252 \\ 0.5 & 506 & -252 \\ 0 & 505.5 & -411 \end{bmatrix} \begin{bmatrix} \frac{\delta a_0}{a_0} \\ \frac{\delta a_1}{a_1} \\ \frac{\delta b_2}{b_2} \end{bmatrix}$$

The cut-off frequency (which may be quite critical) is dependent upon the values of n_0 and n_2, so let us specify an accuracy of 2% for these; m_1 mainly determines the damping factor, the value of which will not usually be crucial, so let us allow 10% for this.

Once again, assuming the same fractional accuracy for all the discrete coefficients, we can divide each continuous coefficient's accuracy by the rss of its sensitivity factors to give the following requirement for the discrete coefficient accuracies:

2% accuracy in n_0 requires 0.0035% accuracy

2% accuracy in n_2 requires 0.0035% accuracy

10% accuracy in m_1 requires 0.015% accuracy

The need to obtain 2% accuracy in n_0 and n_2 is dominant, and thus yields the following values and tolerances for the discrete coefficients:

$$a_0 = 0.996871 \pm 3.5 \times 10^{-5}$$

$$a_1 = -1.989805 \pm 7 \times 10^{-5}$$

$$b_2 = 0.9937412 \pm 3.5 \times 10^{-5}$$

The corresponding numbers of fractional bits can readily be shown to be 14, 13 and 14 bits respectively, giving nearest values as follows:

$$a_0 = 0.996887207 \qquad (0.11111111001101_B)$$

$$a_1 = -1.989746094 \qquad (-1.1111110101100_B)$$

$b_2 = 0.993774414 \qquad (0.11111110011001_B)$

(also $a_2 = a_0$, $b_1 = a_1$)

As before, the process may be completed by determining the implemented values of the equivalent continuous filter coefficients:

$n_0 = 1.0000$ (0% error)

$n_2 = 0.0247$ (2.3% error)

$m_1 = 0.01546$ (2.8% error)

Example 5.3: A notch filter, z-domain design

It is interesting to consider a narrow-band notch filter designed directly in the z-domain, and see how its sensitivity compares with that of the emulated filter which has just been analysed. Figure 5.5 shows the placing of poles and zeros for a second-order filter to achieve the same cut-off frequency and bandwidth as before, and yields the discrete transfer function, for $T = 0.01$ sec:

$$H(z) = \frac{1 - 2\cos\theta z^{-1} + z^{-2}}{1 - 2R\cos\theta z^{-1} + R^2 z^{-2}}$$

in which $\theta = \omega_0 T$ (to give the cut-off frequency)
and $R = 0.9969$ (to give the bandwidth)
that is zeros at $z = 1 \angle \pm\omega_0 T$
poles at $z = 0.9969 \angle \pm\omega_0 T$.

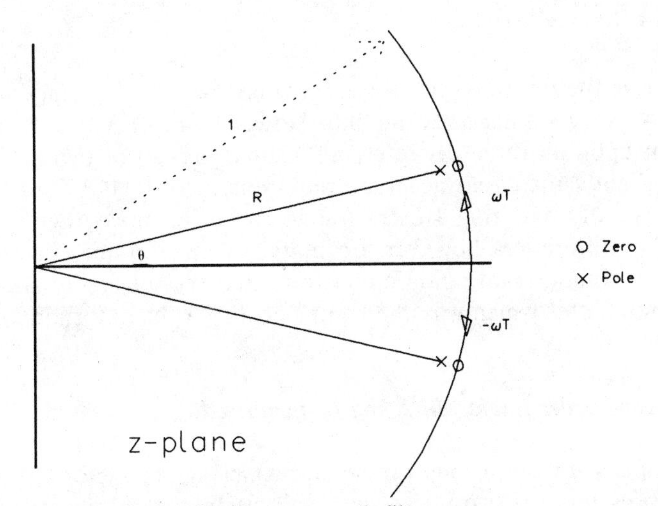

Figure 5.5 Pole placing for discrete notch filter

Thus

$$H(z) = \frac{1 - 1.9960535z^{-1} + z^{-2}}{1 - 1.98987z^{-1} + 0.99382z^{-2}}$$

The rationale for this design is fairly standard [Og87], and is not repeated here, although the principles were outlined in Chapter 3.

Although the denominator coefficients are very similar, the numerator coefficients have changed noticeably. Coefficients a_0 and a_2 are still the same (and equal to 1, which is very easy), but the exact 'pairing' between a_1 and b_1 has disappeared.

Using the emulation table gives equivalent continuous filter coefficients as follows:

$$n_0 = 1.0016$$

$$n_1 = 0$$

$$n_2 = 0.02536$$

$$m_1 = 0.01571$$

$$m_2 = 0.02528$$

A direct comparison of all the sensitivity factors is not possible since the coefficient pairings are different; we therefore restrict the analysis to coefficients which can be compared. Setting $a_0 = a_2 = 1$ gives

$$\frac{\delta n_0}{n_0} = -507.6\frac{\delta a_1}{a_1} + 505.8\frac{\delta b_1}{b_1} - 253\frac{\delta b_2}{b_2}$$

$$\frac{\delta m_1}{m_1} = 0\frac{\delta a_1}{a_1} + 505.1\frac{\delta b_1}{b_1} - 412.8\frac{\delta b_2}{b_2}$$

Comparing these results with those previously derived for the emulated filter shows very similar-sized factors. However the fact that n_0 is now dependent upon all three discrete coefficients, instead of two as before, means a slightly increased accuracy requirement of 0.0026% as against 0.0035%, strictly requiring 1 extra fractional bit in the coefficient word-length. The differences however are marginal in terms of the required precision of computation, and demonstrate that the sensitivity analysis is effective no matter which way the compensator has been formulated.

5.6.4 Second-order filters which can be factorised

If the compensator is a second-order filter which can be factorised, then it can be represented by two first-order sections. In general this will need a lower precision of computation, because it was observed previously that the worst sensitivity factors in a second-order filter are proportional to $1/T^2$, as opposed to $1/T$ for first-order filters.

Consider the following discrete transfer function, which is a 1 Hz low pass filter with 100% damping to give repeated real roots (in s or z):

$$H(z) = 0.000929 \frac{(1 + 2z^{-1} + z^{-2})}{1 - 1.87807z^{-1} + 0.88178z^{-2}}$$

The sensitivity matrix for a second-order section, not repeated here, includes a number of factors in the range 250–500, and requires a coefficient wordlength of 13 bits to obtain 5% accuracy.

The second-order transfer function can readily be factorised into two equal first-order sections:

$$H_1(z) = 0.03047 \frac{1 + z^{-1}}{1 - 0.9390z^{-1}}$$

A sensitivity analysis yields factors no greater than 15, and so in order to achieve the same 5% accuracy requires coefficient wordlengths of 8 bits, which is clearly a significant reduction in precision compared with the single second-order section. In addition, it emerges that the internal wordlength requirements, considered in section 5.7, are also reduced. It would of course require two first-order sections, and examination of the respective equations shows that an extra multiplication is needed, so it is probably only worth splitting into the two first-order sections if very long coefficient wordlengths are being called for. Nevertheless the principle is clear, and certainly if functions higher than second order become necessary, it will almost always be preferable to convert to a combination of first- and second-order sections.

5.6.5 Summary of coefficient accuracy requirements

The foregoing analyses have been based upon specifying a single wordlength requirement to allow for all the coefficients in a particular discrete transfer function. In some cases, different coefficient wordlengths are possible, but this means that a single equation might have a variety of precisions in its computation (feasible but cumbersome). The consistent wordlength means that they will all have the same number of significant bits, although the position of the binary point will move depending upon the coefficient value; the effect of this is easily allowed for by multiplying or dividing by factors of 2 (left or right shifts respectively in binary arithmetic), and will be dealt with in the software section 6.2.

Notice that the total number of bits required to express a coefficient to given fractional accuracy is in practice slightly variable depending upon its actual value. For example, the number 7.9 expressed to 1% accuracy would need 3 integer bits plus 3 fractional bits (giving an accuracy to within $1/2^4$ or ± 0.0625), whereas the number 4.1 expressed to the same accuracy would still need 3 integer bits, but 4 fractional bits are needed because ± 0.0625 is not now sufficient.

Our suggested approach of designing to a fixed coefficient wordlength is probably the most logical. Other approaches are possible however, and the sensitivity matrix could equally well be used to give a fixed number of fractional bits for example.

5.7 The internal variables

The input/output variables, and any intermediate variables within the controller, have been defined to have a certain wordlength from overall system considerations. This will form the basis for the internal variables within the discrete filter sections, but they must also be examined in order to determine any overflow and underflow requirements. Although it is not inevitable (or even obvious) that either or both of these requirements are needed, it is nevertheless essential to check and allow for any such requirements, or else the filter will not work properly under all conditions.

Figure 5.6 summarises a general format for the variables and coefficients. The position of the binary point for the coefficient is predetermined, although the position *shown* for the point in the variables is arbitrary, and is chosen for convenience to make the input/output variables into integers. Overflow allows for 'growth' of the internal variables beyond the maximum size of the input and output variables, and underflow provides fractional bits.

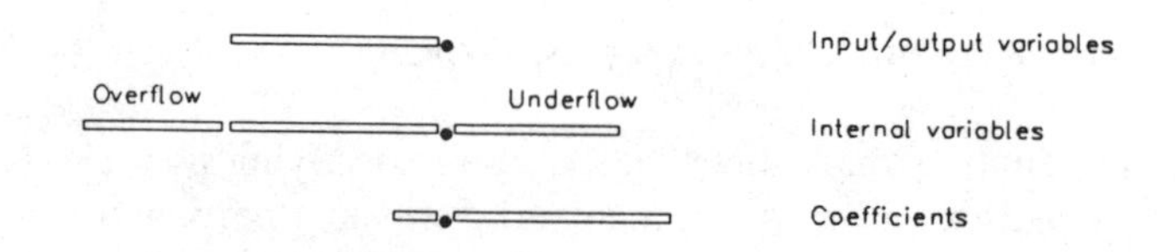

Figure 5.6 General number format

5.7.1 *Internal variable overflow*

A fact that is not widely recognised, but which must be appreciated for successful implementation, is that the internal variables within recursive filters can increase to many times the size of the input, even though the final output may be similar in size to the input. We are primarily concerned here with identifying this increase in the size of the internal variables. Whether or not the output itself will exceed the input depends upon the value of its high- or low-frequency gain, and this issue is dealt with in section 5.7.2.

It is necessary to distinguish between the idea of allowing extra bits for overflow, and that of applying an overflow limit in order to restrict the variable size to its maximum positive or negative value as appropriate. Lack of the latter can cause numerical problems if an overflow occurs, but insufficient room for overflow may cause much more fundamental performance problems, even though a limit is applied. This section mainly deals with allowing sufficient room for overflow, but the following comments about overflow limiting are clearly significant here.

If an internal signal exceeds its maximum range in either an analogue or a digital filter, clearly the filter performance will be affected. Under such conditions, an analogue operational amplifier will inherently saturate at its voltage limits, and normal operation will usually be restored as soon as the saturation condition disappears. Overflow of a signed variable in a digital processor however has a much more profound effect. The two's complement representation of numbers, arbitrarily shown in figure 5.7 for 5 bit numbers, illustrates that an increasingly large positive number can suddenly become a large negative number (or *vice versa*) – restoration of normal operation is by no means assured. Clearly, overflow checks can be made to avoid crossing the boundary, but these are not trivial and can

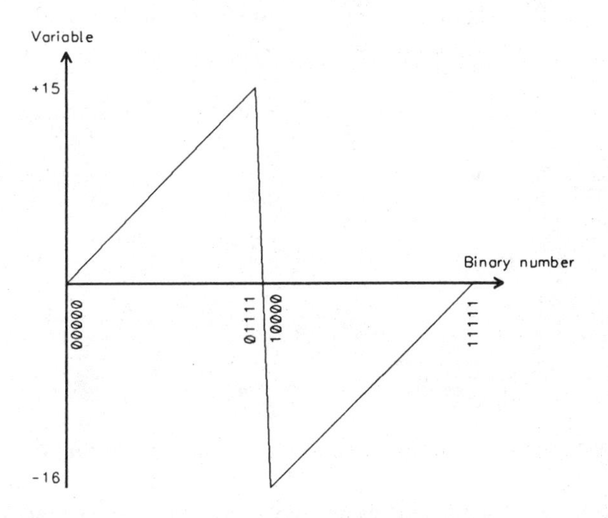

Figure 5.7 2's complement numbers

result in a significant software overhead. A certain amount of overflow checking is inevitable, but it is wise to specify the variable sizes in order to minimise the amount which needs to be done. The software implications of this are dealt with in section 6.2.

To allow for overflow, the maximum size of the internal variable must be determined. The discrete transfer function relating the internal variable v to the input x for a second-order filter was given in section 5.4.2:

$$\frac{v(z)}{x(z)} = \frac{1}{1 + b_1 z^{-1} + b_2 z^{-2}}$$

The zero frequency gain of $v(z)/x(z)$ can be found by substituting $z = 1$ (see section 3.1.2), that is

$$\left.\frac{v}{x}\right|_{ss} = \frac{1}{1 + b_1 + b_2}$$

It is easy to show that for second-order filters with real roots (and, of course, for first-order filters), this ratio is sufficient to define the overflow in v. However for filters having complex roots there is the extra possibility of transient overshoot, particularly with low-damped filters, and so it is sensible to allow an extra factor of 2 (that is, another overflow bit), bearing in mind the comments above about the implications of unchecked overflow. This extra bit will allow for overshoot in most situations, but would be exceeded in the case of a large sustained periodic input at the resonant frequency of a low-damped filter. Such an event is difficult to conceive of in a control system, although in a more general signal processing application it may need to be considered.

The following two examples illustrate how an appropriate allowance for overflow can be determined in specific cases.

Example 5.4 The phase advance compensator used in Example 5.1

For $T = 0.01$ sec, this was:

$$H(z) = \frac{4.8095 - 4.7143z^{-1}}{1 - 0.9048z^{-1}}$$

leading via the canonical realisation to the steady-state relationship:

$$\left.\frac{v}{x}\right|_{ss} = \frac{1}{1 + (-0.9048)} = 10.5$$

No allowance is needed for overshoot in a first-order filter, so 4 overflow bits will be adequate. It must be appreciated that, if these extra bits are not provided, then the compensator will cease to work properly for inputs exceeding about 10% of full scale, even if an overflow limit is applied.

Note that although this has allowed for internal variable overflow, it does not allow for the value of the high-frequency gain of the filter which is greater than unity – this will be discussed in section 5.7.2.

Example 5.5: The notch filter of Example 5.2

In a similar manner to Example 5.4, this leads to:

$$\left.\frac{v}{x}\right|_{ss} = \frac{1}{1 + (-1.989805) + (0.9937412)} = 254$$

Since the filter has a low-damped response, an extra bit should be allowed, making a total of 9 overflow bits. Once again, if these are *not* provided, then problems will be encountered for inputs exceeding approximately 0.4% of full scale – that is, very severe problems!

5.7.2 *Overflow in general*

It is quite fundamental that the overall low-frequency gain through any controller must normally exceed unity once it has been scaled to allow for transducer sensitivities etc. Often quite small percentage errors will generate the maximum output drive signal, and gains ranging from 5 to 20 are typical – sometimes much higher, especially where there is integral action in the controller. The question therefore is how best to cater for this inevitable overflow. The following example will illustrate this.

Example 5.6

The phase advance example used previously would more realistically have a low-frequency gain associated with it, for example:

$$H(s) = 10\left(\frac{1 + 0.5s}{1 + 0.1s}\right)$$

Three approaches are possible, as shown in figure 5.8. Firstly everything can be lumped together, which means that the equations become

$$v_0 = x_0 + 0.9048v_1 \qquad \text{(as before)}$$

$$y_0 = 48.095v_0 - 47.143v_1 \qquad \text{(both coefficients} \times 10\text{)}$$

In the process of computing these equations, firstly an allowance must be made for internal variable overflow (4 bits, as evaluated in section 5.7.1), secondly an *extra* overflow factor of about 50 must be included in the temporary result formed while calculating y (because of the size of the

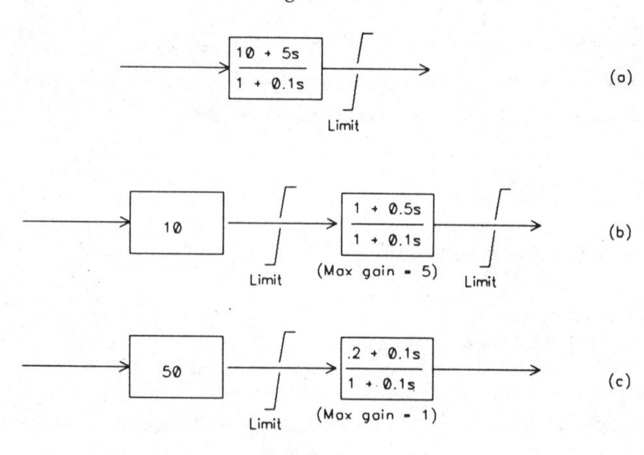

Figure 5.8 Distributing the gain term

coefficients), and thirdly an overflow limit must be applied to the final value of y.

A second approach in which the gain of 10 is applied independently either before or after the phase advance does not help because all the above allowances must be made, *and* an additional overflow limit must be applied when multiplying by 10.

The third approach is to re-organise the compensator into a gain and a filter, with the filter adjusted such that the maximum gain of its frequency response never exceeds unity:

$$H(s) = 50\left(\frac{0.2 + 0.1s}{1 + 0.1s}\right)$$

The gain of 50 will naturally need to be overflow-limited, but the coefficients used in the phase advance now become less than unity. The internal variable overflow must still be allowed for in the same way, but the calculation of the output y can be made without any additional overflow allowance or limit being applied.

The general conclusion therefore is that the third option is best; it is preferable to make all the discrete filter sections have a gain of unity or less, adjusting the associated gain factor as appropriate. A single overflow limit applied when calculating the output from the separate gain factor will then be enough.

Notice that the maximum gain would normally be obvious from the continuous transfer function, whether it occurs at high frequencies as in this example, or at low frequencies such as would be the case with a 'lag' compensator, for example. If however the compensator has been derived directly in the z-domain, the maximum gain may not be as obvious. Substituting $z = 0$ and $z = 1$ will give respectively its instantaneous (that is, high-frequency) and steady-state (that is, low-frequency) gains, so it is still possible to separate out a gain term in order to limit the filter's maximum gain to unity.

5.7.3 *Internal variable underflow*

At this stage, having specified the coefficient accuracy and the overflow factor for the internal variable, we are faced with the situation depicted in figure 5.9 – that is, the multiplication of an integer variable by a coefficient having a certain number of fractional bits. Taken to full precision, the product will contain as many fractional bits as there are in the coefficient. The next time the variable (now containing the fractional bits) is multiplied by the coefficient, it is possible to generate even more fractional bits. This is a recursive process, and so eventually precision must be sacrificed; the variable must be truncated or rounded at some specified precision. So the question is – where do we draw the line? One answer to the question can be generated by a probabilistic approach (see Chapter 4, for example) in which truncation or rounding generates quantisation noise which can be quantified statistically.

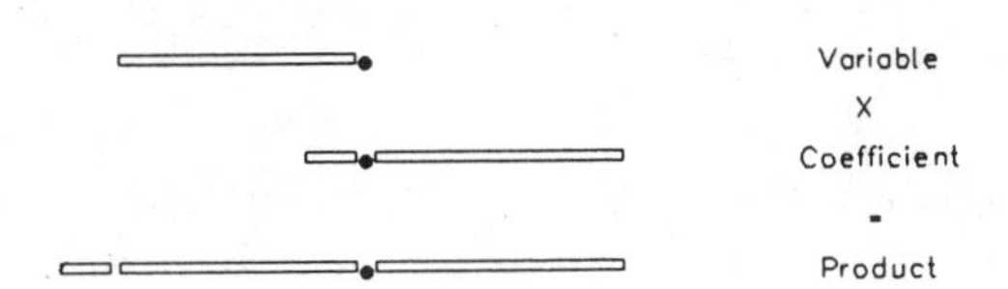

Figure 5.9 Effect of multiplication

It is also possible to pose the problem in a deterministic way. With the input and all internal variables initially set to zero, nothing will happen. If the input then changes from 0 to 1 (that is, the smallest possible change, given the positioning of the binary point proposed in figure 5.6), the output must also change in a manner appropriate to the particular transfer function being implemented. If the output does not respond, then the original specification for the resolution of the input variable has been contravened. (This response to a minimal input was identified as a fundamental requirement in section 5.2). It is intuitively obvious that the problem relates to the number of fractional bits which are provided (that is, to the underflow of the internal variable), and the nature of the question being posed, 'will it respond appropriately, or will it not?', makes it decidedly a deterministic problem.

The equation which is evaluated to give the internal variable v is:

$$v_0 = x - b_1 v_1 - b_2 v_2 \qquad (b_2 = 0 \text{ for first order})$$

When the variable is truncated to i fractional bits, the equation becomes

$$v_0 = \text{integer } \{(x - b_1 v_1 - b_2 v_2)\, 2^i\}/2^i$$

If we now consider the quantised steady-state value v_{ssq} to which v will tend following a minimum sized input (that is, $x = 1$), we can write $v_0 = v_1 = v_2 = v_{ssq}$, and the equation becomes:

$$v_{ssq} = \text{integer } \{(1 - b_1 v_{ssq} - b_2 v_{ssq})\, 2^i\}/2^i$$

Since v_{ssq} is already truncated (that is, $v_{ssq} \times 2^i$ is an integer), it is possible to subtract it from both sides and move it within the integer function without upsetting the equality, giving:

$$0 = \text{integer } \{(1 - v_{ssq}\,(1 + b_1 + b_2))2^i\}/2^i$$

This can be re-expressed by the inequality:

$$0 \leqslant (1 - v_{ssq}\,(1 + b_1 + b_2))2^i < 1$$

since if this expression lies between 0 and 1 its integer portion must be zero. Hence the following range of values for v_{ssq}:

$$\frac{1}{1 + b_1 + b_2} \geqslant v_{ssq} > \frac{1 - 2^{-i}}{1 + b_1 + b_2}$$

A similar calculation, but using a rounded rather than a truncated result, gives:

$$\frac{1 + 2^{-(i+1)}}{1 + b_1 + b_2} \geqslant v_{ssq} > \frac{1 - 2^{-(i+1)}}{1 + b_1 + b_2}$$

The exact steady-state value (that is, without quantisation effects) for an input $x = 1$ would be

$$v_{ss} = 1/(1 + b_1 + b_2)$$

and so the range of possible quantised steady-state values can be written as:

$$v_{ss} \geqslant v_{ssq} > (1 - 2^{-i})v_{ss} \qquad \text{(truncated)}$$

$$v_{ss}(1 + 2^{-(i+1)}) \geqslant v_{ssq} > v_{ss}(1 - 2^{-(i+1)}) \qquad \text{(rounded)}$$

In practice this means that, following the application of a step input $x = 1$ (with the internal variables initially set to zero), the value of the v will increase until it falls within the specified range, and will then 'stick' at a value somewhat short of the exact or unquantised steady-state value. The difference of course represents the error, and the algebra can be adapted to give an expression for the maximum fractional error $\hat{e}_q$ due to quantisation:

$$\hat{e}_q = \left.\frac{v_{ss} - v_{ssq}}{v_{ss}}\right|_{max} = 2^{-i} \qquad \text{(truncated)}$$

$$\text{or } \hat{e}_q = 2^{-(i+1)} \qquad \text{(rounded)}$$

Notice that, in the steady state, the equation by which y is calculated simplifies to

$$y_{ssq} = (a_0 + a_1 + a_2)v_{ssq}$$

and so the fractional error in y will be the same as the fractional error in v_{ssq}, as given above.

The designer can use these error expressions in either of two ways. Firstly the percentage accuracy for the controller performance could be applied, giving fractional bit requirements for our first- and second-order examples as follows:

	Truncated	*Rounded*
Phase advance (5%)	5 bits	4 bits
Notch filter (2%)	7 bits	6 bits

The alternative is to relate the internal variable to what would happen to the output y. For the notch filter example, its unity low-frequency gain means that an input of 1 should eventually produce an output of 1. On the assumption that the actual output for the filter will be the rounded result from the calculation of y, as long as y_{ssq} exceeds 50% of y_{ss} then a rounded output of 1 will (just) be generated. This then gives the fundamental requirement which ensures that a minimal change at the input will cause a corresponding change of the output:

$$\hat{e}_q = 2^{-i} < 0.5 \rightarrow i = 2 \qquad \text{(truncated)}$$

$$\text{or} \rightarrow i = 1 \qquad \text{(rounded)}$$

The two approaches for using the expression for $\hat{e}_q$ represent extremes: less than the minimum would be inadvisable, but more than the maximum is unnecessary. Essentially the underflow analysis gives a deterministic value for the error caused by variable quantisation, and the designer can use it in whatever way is deemed appropriate.

Because it is problematic to specify an appropriate percentage accuracy for a filter's response to the smallest possible input change, these expressions for fractional error due to quantisation can at best be a guide, although experience has shown them to be very useful as such. In the end it is the filter's response dynamically as well as in the steady state which is important, and so a simulation may be the final answer. We recommend that simulation is used, especially for your first few designs, and in order to help we have included a simple but complete program in C for a second-order filter – see Appendix G – this includes the effects of quantisation to a specified number of fractional bits. As the designer's experience grows, it will usually be possible to dispense with the simulation stage in many circumstances.

5.8 Dealing with integrators

The generalised first- and second-order filter sections which have been discussed so far can always be re-organised by separating out a fixed gain so that their maximum gains are always unity, as outlined in section 5.7.2. This means that, as long as an appropriate allowance is made for overflow of the internal variables, overflow detection and limiting of the computations within the digital filters are not necessary; only the gain term will need this. However, the output of a discrete integrator does not have a well-defined maximum or minimum since a sustained input will cause an ever-increasing output; clearly therefore, just as with an analogue integrator, it must be limited to some appropriate level. If a particular compensation stage involves a number of factors including an integrator, it is better to deal with the integrator factor independently.

In this section we discuss the form of a discrete integrator, how to identify its presence in a more complicated discrete compensator, and a method for 'separating it out'. We then elaborate on its implementation and describe the most appropriate method for limiting its output. We have used a specific example to illustrate the principles, which can readily be applied to different situations.

5.8.1 Discrete transfer functions with integrators

Whatever method is used to arrive at a z-domain expression for the integrator, the discrete transfer function will contain a factor $(1 - z^{-1})$ in the denominator, that is, a pole at $z = 1$. Remember, however, that for high sample rates all the poles of a transfer function will be close to unity, but only a pole at exactly unity gives a true integrator. This is another good reason for separate implementation of the integrating factor, a point which will be highlighted in the example which follows. (If the presence of an integrator should be missed for any reason, it will be picked up when determining the internal variable wordlength because there will be an overflow factor of infinity.)

The $(1 - z^{-1})$ factor can now be separated out from the compensator so that it may be implemented separately, and it may be that this leaves a discrete transfer function with a numerator of higher order (in z^{-1}) than the denominator. Unlike *s*-domain transfer functions this does not preclude practical implementation, although it is easy to show that the direct rather than the canonic form must be used. In such cases, therefore, it is sensible to include one of the numerator factors with the $(1 - z^{-1})$ denominator factor, thereby retaining the advantages of the canonic form.

The general transfer function of an integrator, including a numerator factor, can therefore be written as:

$$\frac{y(z)}{x(z)} = \frac{a_0 + a_1 z^{-1}}{1 - z^{-1}}$$

Figure 5.10 is a diagrammatic representation, and the equations to be implemented are:

$$v_0 = x_0 + v_1 \tag{5.1}$$

$$y_0 = a_0 v_0 + a_1 v_1 \tag{5.2}$$

Eqn (5.1) clearly shows that the internal variable v will increase indefinitely with a sustained input x.

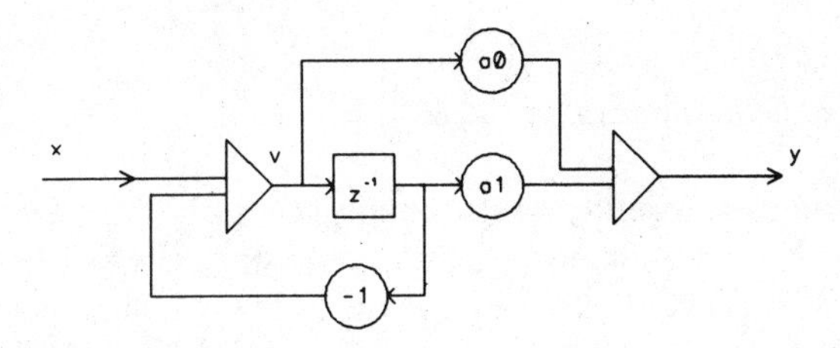

Figure 5.10 *z*-integrator

5.8.2 *Coefficient wordlength*

The denominator coefficient inherently becomes unity – in other words, a multiplication in forming the internal variable v becomes unnecessary, as eqn (5.1) shows. However it is still necessary to determine wordlengths for the numerator coefficients, and the sensitivity factor approach used previously is still applicable.

The generalised discrete transfer function given above can be transformed under the bilinear transform to give an equivalent continuous transfer function:

$$\frac{y(s)}{x(s)} = \frac{1 + n_1 s}{m_1 s}$$

The relationship between the discrete and continuous coefficients can readily be determined:

$$a_0 = \frac{(T + 2n_1)}{2m_1} \qquad n_1 = \frac{T(a_0 - a_1)}{2(a_0 + a_1)}$$

$$a_1 = \frac{(T - 2n_1)}{2m_1} \qquad m_1 = \frac{T}{(a_0 + a_1)}$$

The sensitivity matrix is easily derived as:

$$\begin{bmatrix} \dfrac{\delta n_1}{n_1} \\ \\ \dfrac{\delta m_1}{m_1} \end{bmatrix} = \begin{bmatrix} \dfrac{T^2 - 4n_1^2}{4Tn_1} & \dfrac{4n_1^2 - T^2}{4Tn_1} \\ \\ \dfrac{-(T + 2n_1)}{2T} & \dfrac{2n_1 - T}{2T} \end{bmatrix} \begin{bmatrix} \dfrac{\delta a_0}{a_0} \\ \\ \dfrac{\delta a_1}{a_1} \end{bmatrix}$$

5.8.3 *Overflow of the internal variable*

It is still essential to determine a maximum size for the internal variable v, but in this case it is not dictated by the input, since the smallest possible sustained level will eventually create an overflow. The condition for overflow is now dictated by the output, because it is unnecessary to hold an internal variable which will take the output beyond its limits. Indeed it is distinctly inadvisable to do so, because otherwise this creates the digital equivalent of 'integrator wind-up', a problem which is well known in analogue controllers. This is where the integrator output continues to increase even when the overall output has limited, and means that when the input to the integrator reverses there will be a distinct time delay before the output once again responds to the input (that is, the time during which the integrator ramps down again).

The integral action is of course created by the repetitive addition of the input x to the internal variable v, and it is this process which must be stopped when the output overflows. Hence eqn (5.2) can be used to give

$$y_{max} = (a_0 + a_1)v_{max}$$

from which the 'overflow allowance' can be written as:

$$\frac{1}{a_0 + a_1}$$

This factor can be used in a similar manner to before, but of course it is still necessary to prevent v getting too large, an action which was unnecessary with the non-integrator filter. Procedurally this can be written as, 'If the second equation results in an overflow of y, then do not implement the shift operation' – that is, 'freeze' the integration process by holding v_1 constant. There are other strategies for handling integrator wind-up which are beyond the scope of this book. We have suggested the easiest, but for a more complete discussion, see [Ru90].

Notice that, in the absence of any multiplication when forming the variable v, underflow is not an issue since there is no mechanism for creating underflow bits.

Example 5.7

Consider the discrete compensator

$$H(z) = \frac{3.7583 - 6.65z^{-1} + 2.925z^{-2}}{1 - 1.3333z^{-1} + 0.3333z^{-2}}$$

which is an emulation of a practical PID compensator, having the transfer function:

$$H(s) = \frac{1 + 0.05s}{1 + 0.01s} \cdot \frac{1 + 0.2s}{0.2s}$$

(with a sample frequency of 100 Hz).

Inspection of the denominator coefficients of $H(z)$ should warn of the presence of an integrator, even though the coefficients have been approximated from 4/3 and 1/3. More rigorously, however, factorisation of the discrete transfer function gives:

$$H(z) = 3.7583 \frac{(1 - 0.95122z^{-1})\,(1 - 0.81818z^{-1})}{(1 - z^{-1})\,(1 - 0.3333z^{-1})}$$

This shows the root at $z = 1$ clearly, even though expressed in terms of z^{-1}. This process has also separated out the gain factor of 3.7583.

What we can do therefore is to implement the complete compensator using two sections:

$$H_1(z) = \frac{(1 - 0.81818z^{-1})}{(1 - 0.3333z^{-1})}$$

(which corresponds to the 'phase advance') and

$$H_2(z) = 3.7583 \frac{(1 - 0.95122z^{-1})}{(1 - z^{-1})}$$

(which corresponds to the 'P plus I')

The process of factorisation has limited the maximum gain of the first section to unity, since its instantaneous gain (given by substituting $z^{-1} = 0$) is 1, and its low-frequency gain (given by substituting $z^{-1} = 1$) is 0.273. This means that overflow checks on its output are unnecessary, although an overflow allowance for the internal variable will still be necessary for the section to work properly with large inputs. Implementation of this section is very similar to that described previously.

For the second section it is appropriate to re-combine the gain term with the numerator coefficients, otherwise the objective of preventing the integral action when the output exceeds its limit will not be achieved. This gives the following relationship:

$$y_{max} = (3.7583 - 3.7583 \times 0.95122)v_{max}$$

that is, $a_0 = 3.7583 \qquad a_1 = 3.57497$

giving an overflow allowance of 5.45 for the internal variable (that is, z overflow bits needed).

It is important to contrast this overflow allowance with that made for the internal variables in non-integrator types of filter. In the latter case the allowance made is absolute, because the variables *cannot* normally exceed the given levels, whereas for integrators the internal variables *could* exceed the levels, but do not *need* to be any larger given the maximum size of the output. Additionally, in order to avoid 'integrator wind-up', the summation process which could cause them to become larger needs to be inhibited whenever the output overflows.

The coefficient wordlength also needs to be determined. Its sensitivity matrix works out as:

$$S = \begin{bmatrix} -19.99 & 19.99 \\ -20.5 & 19.5 \end{bmatrix}$$

So, to give an implementation accuracy of 5%, the coefficients would need to be accurate to around 0.18% using the same approach as before for combining the sensitivity factors. This would require 9-bit coefficients with the following actual values:

$a_0 = 11.1100001_B \qquad (3.7578125)$

$a_1 = 11.1001010_B \qquad (-3.578125)$

It is interesting to consider what happens if the same coefficient wordlength is used for the coefficients of the original second-order discrete transfer function (that is, *without* separating out the integrator).

The actual discrete transfer function using 9-bit coefficients becomes

$$H(z) = \frac{3.7578125 - 6.65625z^{-1} + 2.921875z^{-2}}{1 - 1.33203125z^{-1} + 0.333984375z^{-2}}$$

The reverse bilinear transform gives:

$$H(s) = \frac{12 + 4.28s + 0.1707s^2}{1 + 3.41s + 0.034125s^2}$$

It is immediately obvious that the pure integration has disappeared, and factorisation enables it to be re-written as:

$$H(s) = \frac{1 + 0.311s}{0.083 + 0.283s} \cdot \frac{1 + 0.046s}{1 + 0.01s}$$

Comparing this with the original continuous transfer function (from which the discrete transfer function was derived by emulation), it can be seen that the 'phase advance' portion has not been significantly affected by the quantisation of the coefficients. On the other hand, the proportional plus integral part not only has its time constants fairly radically changed, but also it is no longer a true integrator. In other words, in order to guarantee proper implementation of an integrator, it is essential to treat the integrator part separately, in the manner described, because only in this way is there a coefficient of -1 which can be exactly implemented without quantisation error (that is, by dispensing with the multiplication).

5.9 Implementation using the δ-operator

The problems of high coefficient sensitivity, identified by the large sensitivity factors which were calculated in section 5.6, are unavoidable with discrete transfer functions using the z-operator. The long coefficient wordlengths which are a consequence, particularly in high-performance control systems, can impose a significant burden upon the processor.

The so-called delta form was introduced in Chapter 4 as an alternative structure for digital filters. Since its main benefits are realised when it comes to implementation, we are dealing with it in a separate section. The δ-operator, although surprisingly little used, is fundamentally superior to the z-operator because the problems of coefficient sensitivity disappear completely. Much of the theory introduced in the early chapters could have been presented in terms of δ rather than z [Mi90]; however we decided not to do this because of the predominance of the use of z in industry. Nevertheless we recommend that, no matter how the controller is designed, when it comes to implementation the δ-operator should be used.

In the following sections we will analyse the coefficient sensitivity and determine wordlength requirements for the internal variables; many of the calculations will rely on the principles which have been developed for dealing with the z-operator. We will re-design the phase advance and notch filter examples in order to make use of the δ-operator, and draw a comparison with the designs using the z-operator.

5.9.1 The principle of the δ-operator

There are a number of ways in which the δ-operator can be derived. Although it is not the intention to cover these exhaustively, an overview is presented for completeness.

One approach is to inspect the expressions given in tables 5.2 and 5.3 for the denominator coefficients of the functions in z, and observe that they are asymptotic to integer values as T tends to zero (b_1 tends to -1 in first-order filters; in second-order filters, b_1 tends to -2 and b_2 to $+1$). It can then be realised that it is the closeness of the coefficients to these asymptotic values, rather than the values themselves, which determine the characteristics of the response. A solution to the problem of coefficient sensitivity has been attempted [Li71] by re-expressing the coefficients as a combination of their asymptotic values (which can be exactly implemented) and the small differences compared with the asymptotic value (which differences then become asymptotic to zero), as shown in figure 5.11; alternatively, it is possible to create a new operator which has the same effect [Ag75].

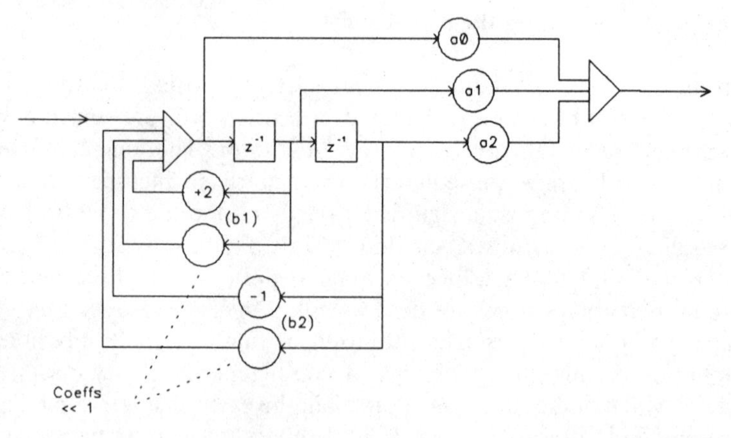

Figure 5.11 Re-arrangement of coefficients

Another approach is to recognise that the linear difference equations which are evaluated are in fact indirectly calculating first and second *derivatives* of the variables by using the first and second differences between successive values of the variables; each coefficient becomes an amalgam of the requirements for a number of these indirect calculations, combined in an equation in such a way as to give the desired dynamic response. It therefore follows that an operator which deals directly in differences between values, rather than in absolute values, must be an improvement.

Either of these approaches leads to the development of the δ-operator [Go85, Goo85], which can be written as:

$$\delta = z - 1$$

$$\delta^{-1} = \frac{z^{-1}}{1 - z^{-1}}$$

5.9.2 *Structures of the δ-filter*

The discrete transfer function in δ can be written in identical form to that for z, although the coefficient values will naturally be different:

$$H(\delta) = \frac{c_0 + c_1\delta^{-1} + c_2\delta^{-2}}{1 + r_1\delta^{-1} + r_2\delta^{-2}}$$

Canonic form

Figure 5.12 is a diagrammatic representation of the δ-filter in canonic form. Notice that independent internal variables v, w and x are used – these are

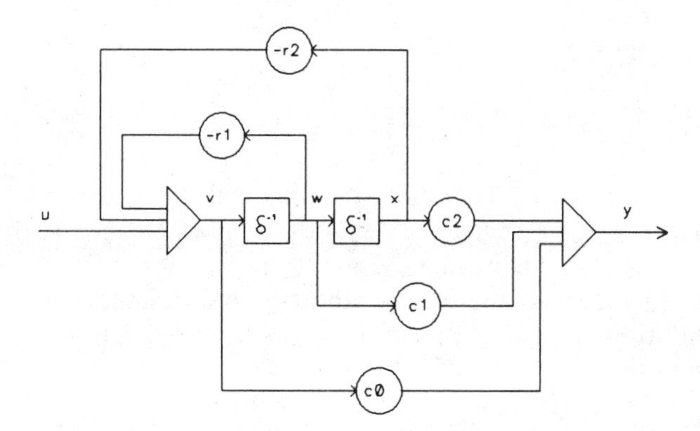

Figure 5.12 Canonic δ-filter

necessary because the δ-filter's internal variables are no longer successive values of the same quantity, and are related by $w = \delta^{-1}v$ and $x = \delta^{-1}w$. As explained in section 3.2, the δ^{-1} operation is an accumulation, meaning that the next value for w is formed by adding v to the current value of w – that is, v is the *difference* between the current and new value of w. The corresponding equations are:

$$v = u - r_1\delta^{-1}v - r_2\delta^{-2}v$$

$$= u - r_1 w - r_2 x$$

$$y = c_0 v + c_1\delta^{-1}v + c_2\delta^{-2}v$$

$$= c_0 v + c_1 w + c_2 x$$

and these transform to:

$$v_0 = u_0 - r_1 w_0 - r_2 x_0$$

$$y_0 = c_0 v_0 + c_1 w_0 + c_2 x_0$$

A C program is again used to list the actual computations which need to be carried out – this best illustrates the real-time processes involved.

```
main( )
{
      while (time < end_of_time)
      {
              if new_sample()
              {
                 u = input_adc();
                 v = u - r1*w - r2*x;
                 y = c0*v + c1*w + c2*x;
                 output_dac(y);
                 /*perform delta operations */
                 x = x + w;
                 w = w + v;
                 time = time + T;
              }
      }
}
```

It is interesting to compare these equations with those given in the C program for the z-filter (section 5.4.2) – the *only* difference is that the original 'shift' equations have been replaced by additions. Otherwise the form of the equations is identical (although the variable names have altered). This simple change results in a fundamental improvement in coefficient sensitivity.

A sensitivity analysis can readily be carried out to show that the sensitivity matrix for such a filter will contain factors which are either close to unity or zero – in other words, the high-sensitivity problems have disappeared, and the percentage accuracy requirements for the coefficients become the same as those required for the filter's performance. We are not including the analysis here because there is an improvement to the structure to discuss, but it is important to realise that it is the introduction of the new operator δ which gives the reduced coefficient sensitivity.

Modified canonic form

Figure 5.13 shows a modification of the filter structure in which the feedback coefficients are moved into the forward path of the filter, with appropriate changes to the coefficients which form the output y. The coefficients are renamed d_1, d_2, p, q and r, to distinguish them clearly from the normal canonic form.

This modification has the important advantage that the internal variables v, w and x now have maximum values which are of the same order as that

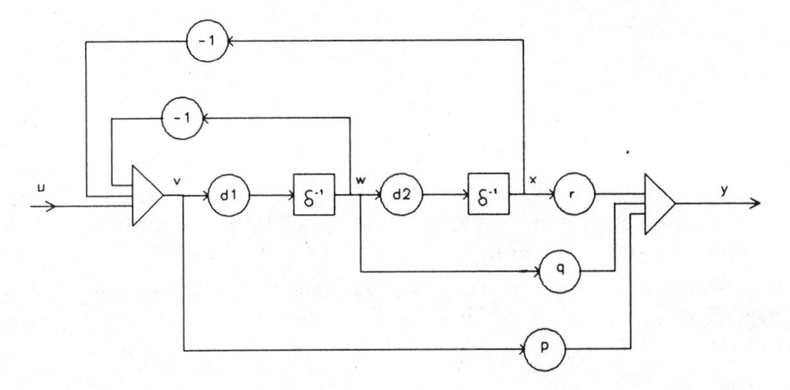

Figure 5.13 Modified canonic δ-filter

of the input variable; in the normal canonic form, they can be shown to be 'magnified' to a size similar to that of the z-filter. This conformity of scaling may be intuitively obvious from the lack of coefficients in the feedback paths of the filter, but the point will be covered further when considering overflow in section 5.9.4. The modification leads to the further advantage that some of the numerator coefficients may be approximated to 0 or 1 in certain circumstances, resulting in a reduction in the number of multiplications needed.

From figure 5.13, the discrete transfer function can be written as:

$$H(\delta) = \frac{p + d_1 q\,\delta^{-1} + d_1 d_2 r\,\delta^{-2}}{1 + d_1\,\delta^{-1} + d_1 d_2\,\delta^{-2}}$$

The corresponding equations are obvious, and transform to:

$$v_0 = u_0 - w_0 - x_0$$

$$y_0 = pv_0 + qw_0 + rx_0$$

The multiplications required for the denominator (that is, those using d_1 and d_2) now appear with the δ^{-1} operations, once again best illustrated by a C program:

```
main()
{
    while (time < end_of_time)
    {
        if new_sample()
        {
            u = input_adc();
            v = u - w - x;
            y = p*v + q*w + r*x;
            output_dac(y);
            /* perform delta operations */
            x = x + d2*w;
            w = w + d1*v;
            time = time + T;
        }
    }
}
```

Tables 5.4 and 5.5 are emulation tables (similar to tables 5.2 and 5.3 for the z-filter) which relate the continuous and discrete filter coefficients for both first- and second-order sections. These can easily be derived using the bilinear transformation expressed in terms of δ, that is:

$$s = \frac{2\delta}{T(2 + \delta)} \quad \text{and} \quad \delta = \frac{sT}{1 - sT/2}$$

Table 5.4 Emulation table for first-order δ-filters

$H(s) = \dfrac{n_0 + n_1 s}{1 + m_1 s}$	$H(\delta) = \dfrac{p + d_1 q \delta^{-1}}{1 + d_1 \delta^{-1}}$
$p = \dfrac{n_0 T + 2n_1}{T + 2m_1}$	$n_0 = q$
$q = n_0$	$n_1 = T\left(\dfrac{p}{d_1} - \dfrac{q}{2}\right)$
$d_1 = \dfrac{2T}{T + 2m_1}$	$m_1 = T\left(\dfrac{1}{d_1} - \dfrac{1}{2}\right)$

Table 5.5 Emulation table for second-order δ-filters

$H(s) = \dfrac{n_0 + n_1 s + n_2 s^2}{1 + m_1 s + m_2 s^2}$	$H(\delta) = \dfrac{p + d_1 q\delta^{-1} + d_1 d_2 r\delta^{-2}}{1 + d_1\delta^{-1} + d_1 d_2\delta^{-2}}$
$p = \dfrac{n_0 T^2 + 2n_1 T + 4n_2}{T^2 + 2m_1 T + 4m_2}$	$n_0 = r$
$q = \dfrac{n_0 T + n_1}{T + m_1}$	$n_1 = T\left(\dfrac{q}{d_2} - r\right)$
$r = n_0$	$n_2 = T^2\left(\dfrac{p}{d_1 d_2} - \dfrac{q}{2d_2} + \dfrac{r}{4}\right)$
$d_1 = \dfrac{4T^2 + 4m_1 T}{T^2 + 2m_1 T + 4m_2}$	$m_1 = T\left(\dfrac{1}{d_2} - 1\right)$
$d_2 = \dfrac{T}{T + m_1}$	$m_2 = T^2\left(\dfrac{1}{d_1 d_2} - \dfrac{1}{2d_2} + \dfrac{1}{4}\right)$

5.9.3 Coefficient wordlength

It has been stated, but not proved, that the δ-operator solves the problem of high coefficient sensitivity. This section will demonstrate that this is so by means of exactly the same procedure which was used for the z-filters. Remember that the emulation tables 5.4 and 5.5 can be used to help the process. Sensitivity analyses are not given in full here, although Appendix F includes the analysis for the first-order example.

Example 5.8: Phase advance compensator

The required continuous transfer function is:

$$H(s) = \frac{G(1 + ks\tau)}{1 + s\tau}$$

Using table 5.4 gives:

$p = G(T + 2k\tau)/(T + 2\tau)$	$G = q$
$q = G$	$k = (2p - d_1 q)/q(2 - d_1)$
$d_1 = 2T/(T + 2\tau)$	$\tau = T\left(\dfrac{1}{d_1} - \dfrac{1}{2}\right)$

and the following sensitivity matrix can be derived:

$$\begin{bmatrix} \dfrac{\delta G}{G} \\ \dfrac{\delta k}{k} \\ \dfrac{\delta \tau}{\tau} \end{bmatrix} = \begin{bmatrix} 0 & 1 & 0 \\ \dfrac{T + 2k\tau}{2k\tau} & -\dfrac{T + 2k\tau}{2k\tau} & \dfrac{(k-1)T}{2k\tau} \\ 0 & 0 & -\dfrac{T + 2\tau}{2\tau} \end{bmatrix} \begin{bmatrix} \dfrac{\delta p}{p} \\ \dfrac{\delta q}{q} \\ \dfrac{\delta d_1}{d_1} \end{bmatrix}$$

Using the same values as before (that is, $G = 1$, $k = 5$, $\tau = 0.1$ with $T = 0.01$) gives the following values for the coefficients and their sensitivities:

$$p = 4.8095$$

$$q = 1$$

$$d_1 = 0.09524$$

$$\begin{bmatrix} \dfrac{\delta G}{G} \\ \dfrac{\delta k}{k} \\ \dfrac{\delta \tau}{\tau} \end{bmatrix} = \begin{bmatrix} 0 & 1 & 0 \\ 1.01 & -1.01 & 0.04 \\ 0 & 0 & -1.05 \end{bmatrix} \begin{bmatrix} \dfrac{\delta p}{p} \\ \dfrac{\delta q}{q} \\ \dfrac{\delta d_1}{d_1} \end{bmatrix}$$

The matrix clearly demonstrates the much reduced coefficient sensitivity compared with the z-filter, with all sensitivity factors either normal or low. Taking all the fractional coefficient accuracies to be the same, and using the rss of the sensitivities as before, in order to achieve a 5% emulation accuracy the worst case would be for k, needing 3.5% accuracy in the coefficients. This can in general be achieved with a 4-bit wordlength (although $q = 1$, of course, will best be implemented without a multiplication):

$$p = 5\ (= 101.0_B)$$

$$q = 1\ (= 1.000_B)$$

$$d_1 = 0.09375\ (= 0.0001100_B)$$

The actual values of the equivalent continuous filter coefficients then become:

$G = 1$ (0% error)
$k = 5.197$ (3.9% error)
$\tau = 0.1017$ (1.7% error)

Notice that d_1 as given above has a wordlength of 7 bits, but only 4 *significant* bits; taking full advantage of the shorter wordlength requires that a scaling is associated with the multiplication, that is, re-writing the coefficient as $0.1100_B \times 2^{-3}$. The exponent is, of course, easily accommodated by means of 'right shifting'; this is discussed in more detail in Chapter 6, but is worth highlighting at this stage.

Example 5.9: Notch filter

The transfer function is:

$$H(s) = \frac{n_0 + n_2 s^2}{1 + m_1 s + m_2 s^2} \qquad \text{(in which } m_2 = n_2\text{)}$$

Using table 5.5 gives:

$$p = \frac{n_0 T^2 + 4n_2}{T^2 + 2m_1 T + 4n_2} \qquad r = n_0$$

$$q = \frac{n_0 T}{T + m_1} \qquad n_2 = T^2 \left(\frac{p}{d_1 q} - \frac{1}{4}\right)$$

$$r = n_0 \qquad m_1 = T\left(\frac{1}{q} - 1\right)$$

$$d_1 = \frac{4T(T + m_1)}{T^2 + 2m_1 T + 4n_2}$$

$$d_2 = \frac{T}{T + m_1} \quad (= q)$$

Note that the above expressions take advantage of paired coefficients.

The usual process can be used to derive a sensitivity matrix as follows:

$$\begin{bmatrix} \dfrac{\delta n_0}{n_0} \\ \\ \dfrac{\delta n_2}{n_2} \\ \\ \dfrac{\delta m_1}{m_1} \end{bmatrix} = \begin{bmatrix} 0 & 0 & 1 & 0 \\ \\ \dfrac{T^2 + 4n_2}{4n_2} & -\dfrac{(T^2 + 4n_2)}{4n_2} & 0 & -\dfrac{(T^2 + 4n_2)}{4n_2} \\ \\ 0 & -\dfrac{(m_1 + T)}{m_1} & 0 & 0 \end{bmatrix} \begin{bmatrix} \dfrac{\delta p}{p} \\ \dfrac{\delta q}{q} \\ \dfrac{\delta r}{r} \\ \dfrac{\delta d_1}{d_1} \end{bmatrix}$$

Taking values as before ($n_0 = 1, n_2 = 0.0253, m_1 = 0.0159$ with $T = 0.01$) gives the values below:

$$\begin{bmatrix} \dfrac{\delta n_0}{n_0} \\ \\ \dfrac{\delta n_2}{n_2} \\ \\ \dfrac{\delta m_1}{m_1} \end{bmatrix} = \begin{bmatrix} 0 & 0 & 1 & 0 \\ \\ 1.001 & -1.001 & 0 & -1.001 \\ \\ 0 & -1.63 & 0 & 0 \end{bmatrix} \begin{bmatrix} \dfrac{\delta p}{p} \\ \dfrac{\delta q}{q} \\ \dfrac{\delta r}{r} \\ \dfrac{\delta d_1}{d_1} \end{bmatrix}$$

The accuracies used when analysing the z-filters give the following requirements:

2% accuracy in n_0 requires 2% accuracy in the coefficients
2% accuracy in n_1 requires 1.15% accuracy in the coefficients
10% accuracy in m_1 requires 6.1% accuracy in the coefficients

The requirement for n_1 is dominant, and calls for coefficients expressed with the following tolerances:

$$\begin{aligned} p &= 0.9969 \pm 0.0115 \\ q &= 0.3861 \pm 0.0044 \qquad (= d_2) \\ r &= 1 \pm 0.0115 \\ d_1 &= 0.01095 \pm 0.00012 \end{aligned}$$

It is straightforward to calculate the number of fractional bits for each, and their nearest values:

$p = 1$	(1.000000_B)	(6 fractional bits)
$q = 0.382815$	(0.0110001_B)	(7 fractional bits)
$r = 1$	(1.000000_B)	(6 fractional bits)
$d_1 = 0.010986$	(0.000000101101_B)	(12 fractional bits)

(and $d_2 = q$)

Coefficients p and r would not actually be implemented of course, since multiplying by 1 is trivial. Although the other three coefficients have differing numbers of fractional bits, each needs six significant bits, combined with an appropriate number of right shifts of the variable (division by 2^1 for q and d_2, and by 2^6 for d_1).

The corresponding values and errors for the equivalent continuous coefficients are:

$n_0 = 1$ (exact)

$n_2 = 0.0256$ (1.1%)

$m_1 = 0.01612$ (1.4%)

In practice, it is often not necessary to carry out the sensitivity analysis for δ-filters. It is worth knowing how many of the discrete coefficients affect each continuous coefficient, in order to allow for this when determining the required accuracy, but this can usually be discovered by inspection of the algebraic expressions for the continuous coefficients.

5.9.4 *Internal variable overflow*

It is necessary to ensure that none of the internal variables either saturate or overflow when the maximum input is applied. As with the z-filters, there may be some low- or high-frequency amplification, in which case saturation is inevitable, but this can be handled by means of a separate gain factor, exactly as for the z-filter. Whereas the internal variables of a z-filter are equal in the steady state, and track each other very closely under dynamic conditions, the internal variables in a δ-filter are quite different, and it is therefore necessary to consider them separately.

Their discrete transfer functions in δ, for a second-order example, are:

$$\frac{v}{u} = \frac{1}{1 + d_1\delta^{-1} + d_1 d_2\delta^{-2}}$$

$$\frac{w}{u} = \frac{d_1\delta^{-1}}{1 + d_1\delta^{-1} + d_1 d_2\delta^{-2}}$$

$$\frac{x}{u} = \frac{d_1 d_2 \delta^{-2}}{1 + d_1\delta^{-1} + d_1 d_2 \delta^{-2}}$$

Remember that initial and final values to a step input $u = u_{max}$ can be determined by substituting $z^{-1} = 0$ and $z^{-1} = 1$, that is, $\delta^{-1} = 0$ and ∞ respectively; this yields the following values for a second-order filter:

	Initial $\delta^{-1} = 0$	*Final* $\delta^{-1} = \infty$
v	u_{max}	0
w	0	0
x	0	u_{max}

Since v starts at u_{max} and then settles to zero, we can infer that there may be some undershoot to a negative value before settling, but that the maximum value will be the initial value, that is, u_{max}. Variable x starts at zero and settles to u_{max}, possibly with overshoot to nearly twice this value in low-damped filters; hence its maximum value will be $2u_{max}$.

The initial and final value analysis is not helpful in determining the maximum value of the variable, w, however. One approach would be a rigorous mathematical analysis of its maximum value under different conditions, but it is as easy to demonstrate by simulation that it has a similar maximum to v and x (this is to be expected, both from the maxima derived for v and x, and from the basic structure of the filter having unity 'feedback' coefficients). Figure 5.14 shows the time responses for w obtained from simulations of a 1-Hz second-order section with various damping factors, and this shows maxima always less than u_{max}. (The graphs

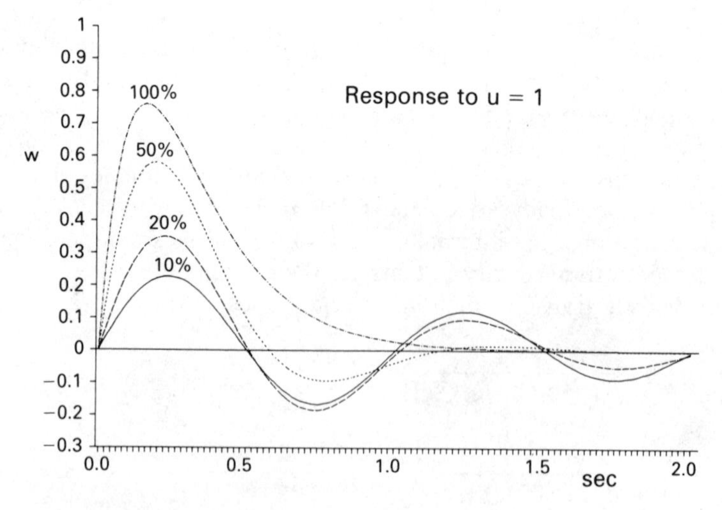

Figure 5.14 Simulation to show size of w

are for a sample rate of 100 Hz, but are substantially unchanged for higher or lower sample rates).

In a first-order filter, the initial and final values are sufficient to determine maximum values, and can readily be shown to be u_{max}; an overshoot allowance is unnecessary.

In conclusion, overflow of the internal variables in the modified canonic form of the δ-filter is not a problem; in some cases it may be necessary to allow an extra bit for overshoot, but generally the internal variables are similar in size to the input and output variables, irrespective of sample rate. It is worth contrasting this with the *z*-filter, in which overflow must be increasingly allowed as the sample period decreases.

5.9.5 *Internal variable underflow*

The independent nature of the three internal variables makes this a little less straightforward than the calculations for the *z*-filter.

Examining the error caused by quantisation in the steady-state response to a step input is an expedient method. Ultimately it is the accuracy of the output, rather than that of the internal variables, which matters, although if the particular filter has a high pass response (that is, a zero steady-state output) it will be necessary to consider the internal variables instead. Consequently, the analysis which follows derives expressions for the fractional error due to quantisation both in the output *and* in the internal variables.

Let us look at the equations:

$$v = u - w - x$$

$$x = x + d_2 w$$

$$w = w + d_1 v$$

$$y = pv + qw + rx$$

(Remember that in the *z*-filter, *only* the internal variables needed to be considered because in steady state the output is dependent upon the single value to which they all converge).

A steady state following a step input is only achieved when the results of $(d_2 w_{ssq})$ and $(d_1 v_{ssq})$ are both zero, since this means that x and w will no longer change, that is:

$$\text{integer } \{d_1 v_{ssq}\, 2^i\}/2^i = 0 \text{ where } i \text{ is the number of fractional bits}$$

and

$$\text{integer } \{d_2 w_{ssq}\, 2^i\}/2^i = 0$$

These can be re-expressed using the inequalities:

$$0 \leqslant d_1 v_{ssq}\, 2^i < 1$$

$$0 < d_2 w_{ssq}\, 2^i < 1$$

or

$$0 \leqslant v_{ssq} < \frac{2^{-i}}{d_1}$$

$$0 \leqslant w_{ssq} < \frac{2^{-i}}{d_2}$$

These expressions indicate the maximum errors in two of the three internal variables. The maximum error in the third variable, x, can be derived as follows. It is possible to write down an expression for the quantised steady-state value x_{ssq} in response to a minimum input $u = 1$ as:

$$x_{ssq} = 1 - w_{ssq} - v_{ssq}$$

Substituting the range of errors in v and w gives:

$$1 \geqslant x_{ssq} > 1 - \frac{2^{-i}}{d_1} - \frac{2^{-i}}{d_2}$$

Since $x_{ss} = 1$, it is possible to derive the maximum fractional error in x due to quantisation:

$$\hat{e}_{xq} = \frac{d_1 + d_2}{d_1 d_2} 2^{-i}$$

Re-worked with rounded rather than truncated calculations gives:

$$\hat{e}_{xq} = \frac{d_1 + d_2}{d_1 d_2} 2^{-(i+1)}$$

Corresponding errors in w for first-order filters can readily be shown to be:

$$\hat{e}_{wq} = \frac{2^{-i}}{d_1} \text{ (truncated)} \qquad \text{or} \qquad \frac{2^{-(i+1)}}{d_1} \text{ (rounded)}$$

This has identified errors in the internal variables, and these can now be used to predict errors in the output y due to quantisation, including its fractional error e_{yq}:

$$\begin{aligned} y_{ssq} &= p v_{ssq} + q w_{ssq} + r x_{ssq} \\ &= p v_{ssq} + q w_{ssq} + r(1 - v_{ssq} - w_{ssq}) \\ &= r + (q - r) w_{ssq} + (p - r) v_{ssq} \end{aligned}$$

that is

$$r \leqslant y_{ssq} < r + (q - r)\frac{2^{-i}}{d_2} + (p - r)\frac{2^{-i}}{d_1}$$

$$0 \leqslant e_{yq} < \left(\frac{q}{r} - 1\right)\frac{2^{-i}}{d_2} + \left(\frac{p}{r} - 1\right)\frac{2^{-i}}{d_1} \quad (\text{since } y_{ss} = r)$$

and

$$\hat{e}_{yq} = \left(\frac{rd_1 - qd_1 + rd_2 - pd_2}{rd_1d_2}\right)2^{-i}$$

A similar calculation for the error in y for a first-order filter gives the following expression:

$$\hat{e}_{yq} = \frac{(p - q)2^{-i}}{qd_1}$$

Applying these results to the filters used in the examples yields fractional bit requirements as follows:

	Truncated	*Rounded*
Phase advance (5%)	8 bits	7 bits
Notch filter (2%)	13 bits	12 bits

The 50% accuracy requirement needed to ensure propagation of the minimum input can be calculated as:

	Truncated	*Rounded*
Phase advance	5 bits	4 bits
Notch filter	8 bits	7 bits

5.9.6 *Integrators using* δ

The need to limit the output of an integrator, and the consequent advisability of treating it separately (see section 5.8), applies equally well to an integrator using δ. Consequently, we re-work the example used for the z-integrator in order to demonstrate the way the equivalent δ-filter needs to be formulated.

Remembering that $\delta = z - 1$, a pole at $z = 1$ becomes a pole at $\delta = 0$; this immediately reveals the condition for a discrete transfer function in δ to include an integrating factor, and may be demonstrated using the general integrator function in z which was given in section 5.8.1:

$$H(z) = \frac{a_0 + a_1z^{-1}}{1 - z^{-1}}$$

Substituting for δ and re-arranging gives:

$$H(\delta) = \frac{a_0\delta + (a_0 + a_1)}{\delta} = c_0 + c_1\delta^{-1}$$

that is

$$c_0 = a_0, \qquad c_1 = a_0 + a_1 \qquad \text{and} \qquad r_1 = 0$$

Having no denominator, of course the corresponding equations are very simple because there is no recursion. Here they are written with a coefficient before the δ^{-1} operation, and the corresponding diagrammatic representation is given in figure 5.15:

$$y(k) = c_0 u(k) + w(k) \tag{5.3}$$

$$w(k + 1) = w(k) + c_1 u(k) \tag{5.4}$$

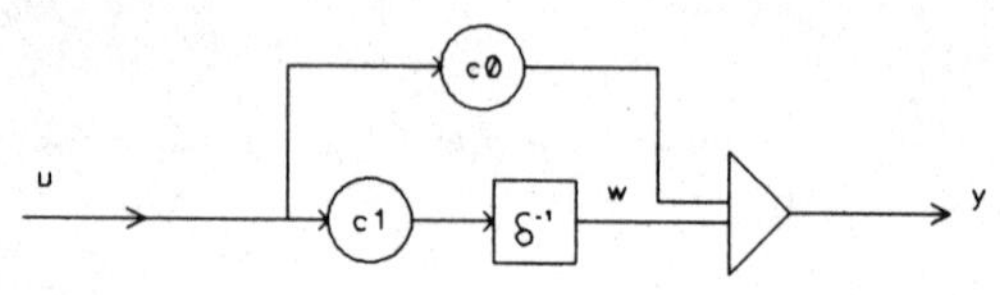

Figure 5.15 δ-integrator

Note that the variable w is used for compatibility with the more general first-order filter in δ, the other variable v being unnecessary.

As with the z-integrator, the maximum size w_{max} needed for the internal variable w is determined by the maximum y_{max} stipulated for the output y. It is clear from eqn (5.3) that

$$y_{max} = c_0 u + w_{max}$$

so, for a given value of y_{max}, w_{max} is greatest when $u = 0$. Thus:

$$y_{max} = w_{max}$$

If during computation the value of y exceeds its limits, variable w must be increased no further.

The coefficient sensitivity, which we have not included, shows 'normal' sensitivity factors (that is, close to unity).

Underflow must be considered, because multiplication by a fractional coefficient in eqn (5.4) will cause fractional bits. The quantisation error from the multiplication in eqn (5.3) will appear directly as an error in y. For a minimum size input $u = 1$, the quantisation error with i fractional bits is

$$e_w = c_1 - \text{integer}\,(c_1 2^i)/2^i \quad \text{(for truncation)}$$

Note that this is the error in the slope of the integrator output in response to an input $x = 1$, which should be c_1 per sample period. It is useful to express this as a *fractional* error in the slope:

$$e_{wf} = \frac{c_1 - \text{integer}\,(c_1 2^i)/2^i}{c_1}$$

$$= \frac{c_1 2^i - \text{integer}\,(c_1 2^i)}{c_1 2^i} \tag{5.5}$$

Observing that the maximum value of the numerator in eqn (5.5) is 1, we can now deduce a maximum value for the fractional error due to quantisation as

$$\hat{e}_{\text{wf}} = \frac{1}{c_1 2^i}$$

Notice that a judicious choice of T could be made, such that the value of c_1 would be expressible in many fewer fractional bits than is predicted by $\hat{e}_{\text{wf}}$. This is not to be recommended however, since it may be necessary to make changes to c_1 for which the reduced number of underflow bits would be insufficient. It is more important to consider carefully what size of fractional error should be designed for, remembering that the internal variable quantisation is only a problem with small inputs.

Example 5.10

Consider the integrator section from the example used in section 5.8, which had the discrete transfer function:

$$H(z) = 3.7583\,\frac{(1 - 0.95122z^{-1})}{(1 - z^{-1})}$$

Substituting $\delta = z - 1$ gives the equivalent discrete transfer function in δ:

$$H(\delta) = 3.7583 + 0.183\delta^{-1}$$

For 5% implementation accuracy, the coefficients both need 4 significant bits, and can readily be worked out to be:

$$c_0 = 3.75 \quad (= 11.11_{\text{B}})$$

$$c_1 = 0.1875\ (= 0.001100_{\text{B}})$$

Working on 5% accuracy for the quantisation error means that

$$0.05 \geqslant \frac{1}{(0.183)2^i}$$

from which $i = 7$ fractional bits.

For interest, table 5.6 lists the actual percentage quantisation error e_{wf} for different numbers of fractional bits, with a range of values for c_1 from 0.001010_{B} through to 0.001110_{B}, that is, two values either side of the quantised coefficient value, giving approximately ±20% variation. This shows that fewer than the 7 fractional bits predicted by $\hat{e}_{\text{wf}}$ are needed in all five cases. There are intermediate values for c_1 which would require the full 7 bits to achieve 5% accuracy, but of course given the coefficient quantisation (also related to the 5% accuracy requirement) these values would never be used. The table shows a subtle relationship between the variable and coefficient quantisation, but it is unfortunately mathematically intractable, so it is difficult to draw any rigorous conclusions. Nevertheless, $\hat{e}_{wf}$ gives a good if pessimistic indication, and e_{wf} can always be explicitly calculated if necessary.

Table 5.6 Percentage errors e_{wf} due to internal variable quantisation

Fractional bits	c_1				
	0.001010_B	0.001011_B	0.001100_B	0.001101_B	0.001110_B
3	20	27.3	33.3	38.5	42.9
4	20	27.3	0	7.7	14.3
5	0	9.1	0	7.7	0
6	0	0	0	0	0
7	0	0	0	0	0

5.10 Summary of the effects of digitisation

What we have done is to identify the effects of converting a discrete transfer function into a digital filter in which both coefficients and variables are subject to the effects of quantisation. We have presented this from a design point of view, so that rather than *analysing* the errors introduced for example by using 8-bit wordlengths, we have given techniques for specifying the computational precision needed to satisfy overall accuracy requirements for the controller. We can now draw together these computational requirements by summarising the variable and coefficient wordlengths for the two compensator examples which have been used throughout this chapter.

Figure 5.16 shows these for both the z- and δ-filters, based upon a 12-bit wordlength for the input and output variables. The coefficient format varies, so just one is shown. The comparison is clear: there is a small difference in the requirements for the internal variables, with those in the δ-filters being slightly shorter, but the major difference comes with the wordlengths for the coefficients, because for the δ-filter there is a very significant reduction. It should also be noted that if the sample frequency were increased, the z-coefficients would need to be even longer, whereas the number of significant bits in the δ-filter coefficients would remain unchanged. There is another implication however: the low coefficient sensitivity of the δ-filter means that some of the coefficients may, with little effect, be approximately to unity, so that the corresponding multiplication

Phase advance network

			bits	*section*	*example*
z	variable	overflow	4	5.7.1	5.4
		basic	12	5.10	
		underflow	5	5.7.3	
	coefficient		11	5.6.2	5.1
δ	variable	overflow	0	5.9.4	
		basic	12	5.10	
		underflow	8	5.9.5	
	coefficient		4	5.9.3	5.8

Notch filter

			bits	*section*	*example*
z	variable	overflow	9	5.7.1	5.5
		basic	12	5.10	
		underflow	7	5.7.3	
	coefficient		14	5.6.3	5.2
δ	variable	overflow	1	5.9.4	
		basic	12	5.10	
		underflow	13	5.9.5	
	coefficient		6	5.9.3	5.9

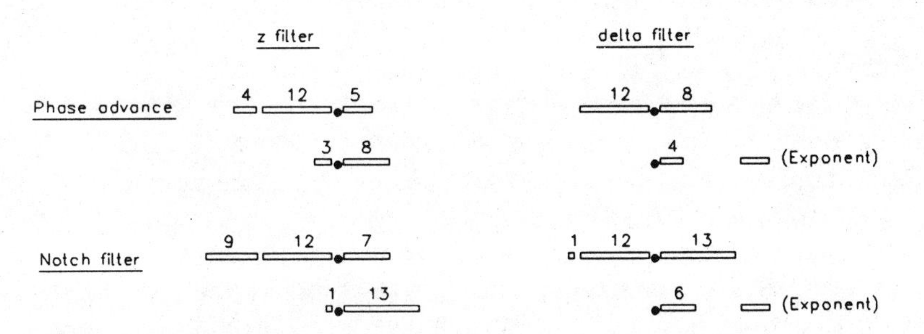

Figure 5.16 Summary of variable and coefficient wordlengths

is unnecessary. For the phase advance compensator, the z-filter requires three 11-bit × 21-bit multiplications, whereas the δ-filter needs only two of 4-bit × 20-bit precision. The notch filter example has a similar saving: five multiplications of 14 bits × 28 bits for the z-filter, and only three multiplications of 6 bits × 26 bits for the δ-filter. This reduction in the number of multiplications is not restricted to these two examples – it is a frequent occurrence with δ-implementation.

What we have described has deliberately ignored aspects relating to specific architectural features of the processor to be used. In practice, of course, the actual wordlength used for both variables and coefficients will be multiples of 8 or 16 bits. The adaption of the basic requirements for computational precision to a specific type of processor will be dealt with in the following chapter.

For the reader's convenience, the examples used in this chapter are summarised as follows:

Filter	*Topic*	*Domain*	*Page*	*Example*
Phase advance	Coefficient accuracy	z	127	1
Notch – emulated design	Coefficient accuracy	z	132	2
Notch – z-plane design	Coefficient accuracy	z	135	3
Phase advance	Overflow	z	140	4
Notch	Overflow	z	141	5
Phase advance with lf gain	Overflow	z	141	6
PID example	General design	z	149	7
Phase advance	Coefficient accuracy	δ	157	8
Notch	Coefficient accuracy	δ	159	9
Integrator	General design	δ	167	10

6 Specific Issues of Implementation

The previous chapter examined the general issues of implementation, primarily the conversion of discrete compensator transfer functions into equations, and the determination of what computational precision is needed within those equations. The next stage is to translate these requirements into a specific combination of hardware and software which will implement the controller in an accurate, effective and efficient manner. This chapter presents an overview of the techniques needed to do this, although it is not necessary to be comprehensive because of the many books dealing in detail with all aspects of microprocessor systems.

The hardware and software considerations are dealt with separately, although in nearly all cases there will be a strong interaction between the two. Such interaction occurs not only at the stage of deciding upon the particular processor to be used, but also when it comes to the more detailed design aspects.

6.1 Hardware for digital controllers

The design of digital circuits incorporating processors is a complex procedure well described in other texts, and it would be inappropriate to cover all the issues in a book of this sort. Nevertheless the hardware aspects are clearly an essential ingredient of the implementation process, and so a review of the major issues is appropriate. The intention is to highlight the aspects likely to be of greatest significance in a control type of application, but specifically *without* giving all the design information which would be needed in practice. By this approach we hope to give a sufficient feeling for the hardware implications which, combined with the software considerations a little later in the chapter, will give a reasonably balanced perspective of the whole subject of digital control in practice. We discuss categories of control hardware and the types of processor which may be used, outline the principles for interfacing digital and analogue input/output signals, and present some examples of design.

The choice of hardware will depend very much upon the application, and also upon the expertise and corporate profile of the company making the control system. Broadly the possibilities may be subdivided into three categories:

Mini or microcomputer
General-purpose single board computer system
Dedicated or 'embedded' controllers

In the first category, a normal computer is adapted by means of plug-in cards to provide the interface requirements such as analogue-to-digital and digital-to-analogue conversion. It will be programmed in some suitable manner to carry out the necessary control functions; once the control program is running the keyboard can be used to interact with its operation, and the monitor may display information relating to how the system is performing. The constructional standard of the computer for an office or laboratory may be unsuitable for a harsher industrial control environment, but 'ruggedised' rack-mounted versions of some personal computers are available to meet this need.

The second option is where a standard design of microprocessor system is sold as a designed-and-tested printed circuit board for incorporation into a particular application. Such products often include all the facilities needed, but may be expanded by means of a standard bus or interconnection structure to incorporate extra facilities as necessary. In many circumstances these 'single board computers' will include a serial communications link to a terminal such that the computer can be programmed. They usually include a simple Machine Operating System – that is, a piece of software, probably held in EPROM, which provides drivers for the terminal and a simple set of low-level commands for examining memory, setting breakpoints in the software, etc. Program development can be carried out on a separate computer, transferred either by the serial link into the memory or by means of EPROM. Sometimes a mass storage device such as a floppy disk (or even a hard disk nowadays) may be added, which means that compilers and assemblers can be run within the single board computer itself, although this may also require that the memory is expanded.

The third category is a dedicated control circuit designed-to-purpose with just what is needed in the way of memory and interfacing, and built-in or embedded within the complete system which is to be controlled. The processor will normally be programmed on a separate Microprocessor Development System, possibly using facilities such as In Circuit Emulation (ICE) for program development, with the final program residing in EPROM. A serial communication link to a terminal may be incorporated, but will usually be for simple interactive tasks only, such as altering a few parameters, although even this may be an unnecessary luxury in applications involving significant production quantities.

In the sections which follow, the emphasis is on the requirements for the third category, in which the hardware is to be designed to purpose, rather than 'bought-in'. However, the issues which are discussed have clear relevance to the other two categories because it will be necessary to understand what is available in order to specify what is needed, and hence select a suitable product on the basis of the requirements.

6.1.1 Types of processor

The intention here is not to stray into the detail of particular processors (of which there are many), but rather to discuss generic types; the distinction between them will help the reader to appreciate the most important hardware implications of digital controllers. Although within each generic type we include a particular example drawn from our own experience, the choice of a specific microprocessor to meet the requirements of a particular application is a complex process and will rely upon more than just technical issues. Considerations such as company expertise, the existing development support, second sourcing, etc. may become the deciding factors in a practical development situation.

(a) Basic microprocessors

The essence of a microprocessor is depicted in figure 6.1. There will be a number of local storage locations (usually called registers) and an arithmetic logic unit (ALU). The arithmetical operations and their operands are specified by a sequence of instructions held in an instruction register. Various control signals are essential, primarily for timing but also to create functions such as interrupts. Communication to and from the processor will normally be via address and data busses, the latter carrying not only numerical information to be processed but also program code fetched from ROM or EPROM and held in the instruction register during execution.

The basic microprocessor cannot operate on its own: it needs to be connected to other integrated circuits, in particular memory devices for storing data and programming information. If timing functions are needed, a timer/counter must be added; if digital input/output is necessary, then some sort of input/output adaptor must be added – all these external devices will be accessed via the processor's busses. A basic microprocessor is extremely versatile because whatever is necessary can be added – as little or as much as is needed. It does however mean that a minimum system must consist of the processor itself, some RAM and EPROM, probably at least one timer/counter, and normally some analogue and digital input/output.

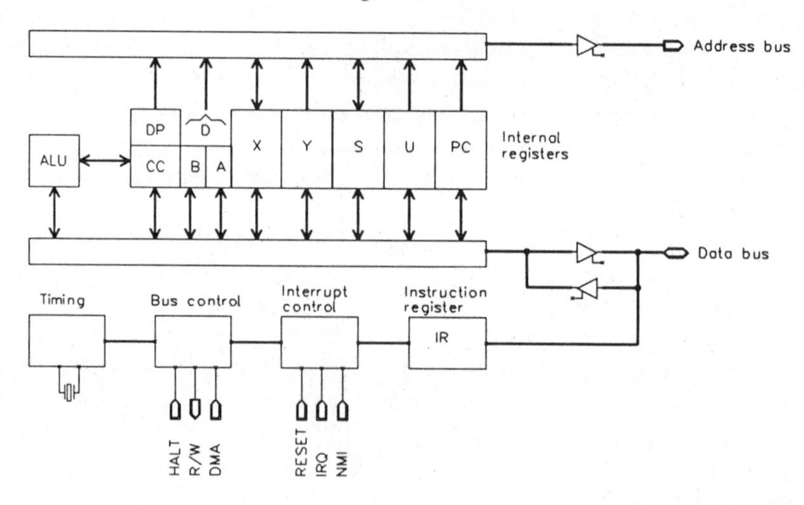

Figure 6.1 Microprocessor architecture

Since the devices themselves are so versatile, their instruction sets must be sufficiently general purpose to satisfy the requirements of all the applications to which they may be put.

A variety of 8-, 16- and 32-bit microprocessors is available. Some of the early 8-bit devices could only perform basic functions such as addition, subtraction and Boolean operations, with very restrictive addressing modes; more modern 8-bit processors include a multiply instruction and more sophisticated addressing modes. Virtually all 16- and 32-bit processors include multiplication, some with floating-point capability, and many of the newer devices are capable of accessing large amounts of external memory in a variety of often quite complex addressing modes, these features being necessary for modern computer systems.

(b) Single-chip microcontrollers

The microprocessor manufacturers recognised some time ago that a reduction in system complexity primarily depended upon having fewer integrated circuits, and for this reason a number of devices are now available in which some of the essential functions are included with the processor in a single package (see figure 6.2). Since such devices can operate on their own, they are sometimes called single-component 'microcomputers', but the more popular term nowadays is ' microcontrollers' since this better reflects the most likely purpose to which they will be put.

In addition to the processor itself and its associated control, all such devices have data memory (RAM), timer/counters and programmable

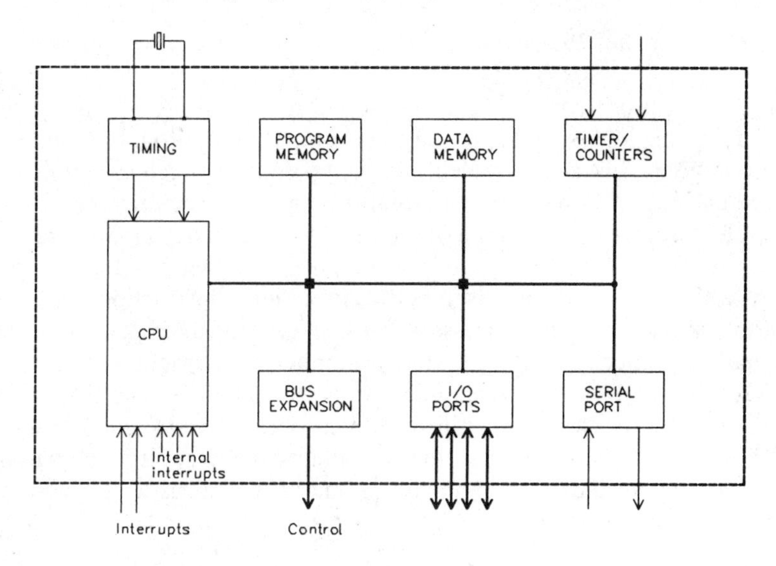

Figure 6.2 Single-chip microcontroller architecture

input/output 'ports'. Nearly all have the facility for including program memory, either in the form of 'mask-programmable' ROM for quantity applications (programmed to a customer's requirement at the time of manufacture), or EPROM for smaller-quantity applications and for development purposes. Usually the devices are offered as a family, different members of the family having extra in-built facilities, such as A/D conversion, serial input/output, etc. Most offer the capability for expansion to include external memory devices if required, although this detracts from the single component concept. This expansion may also inhibit the use of some of the microcontroller's pin connections for control purposes, since external bus expansion is usually through the input/output connections.

The instruction sets of microcontrollers are in general more restricted than the microprocessor, and memory may often be quite limited, both in the amount provided and in the modes of addressing. Often instructions are provided which permit individual manipulation of the device's input/output lines, clearly a most useful feature in many control applications.

There is a reasonable number of 8-bit microcontrollers, and a somewhat smaller number of 16-bit devices. The 8-bit families range from first-generation devices with a very simple set of instructions through to second-generation devices with higher speed and greater arithmetic capability (including multiplication). Generally, in the region of 100–200 bytes of RAM and 4K bytes of ROM or EPROM are available on the chip.

(c) Digital Signal Processors (DSPs)

This type of processor has been developed for signal processing applications requiring high-speed and real-time computation, often for communication systems, but also applicable to digital control. Architecturally they are often quite different from normal microprocessors, and they have a separate 'program' bus for instructions (the so-called Harvard architecture – see figure 6.3). For speed, the ALU will have a hardware multiplier array which can carry out its function in a single instruction cycle (the multiplication in microprocessors and microcontrollers, if any, will normally be 'micro-coded', and take a number of instruction cycles).

DSPs fall somewhat between microprocessors and microcontrollers in a functional sense – a basic DSP will incorporate RAM and ROM or EPROM, but still require the other peripheral devices, although variants are becoming available which include some of the extras which would be expected as standard in a microcontroller.

The instruction set is targeted towards the needs of high-speed computation. Although the basic multiplication is very fast, increased precision by means of software routines may be cumbersome. Certainly the range of instructions available in a normal microprocessor will not be available here, and this may make some of the more mundane operations such as interfacing to analogue input/output or other peripheral devices clumsy. Nevertheless these processors provide an attractive option when a high complexity of control is needed, simply because of the speed at which they carry out the computation.

DSPs start at 16 bits, with 24- and 32-bit devices also available. The latter may include a floating-point CPU, but are aimed towards very high performance computer workstations rather than embedded controllers.

(d) Miscellaneous

For completeness, two more types are mentioned. The first is a semi- or full-custom integrated circuit designed for a specific application or range of applications (more of a design strategy than a generic type). Clearly, in this case the development time and cost will normally preclude small-quantity applications, but the approach offers the chance to 'streamline' the processor's design to the application's specific computational needs. The second type is the transputer, of particular interest for very sophisticated control strategies in which a high degree of parallelism may be necessary in computational requirements. Such controllers are beyond the remit of a text covering the fundamentals, but as devices such as the transputer become commercially more mature it may well be that they are increasingly used.

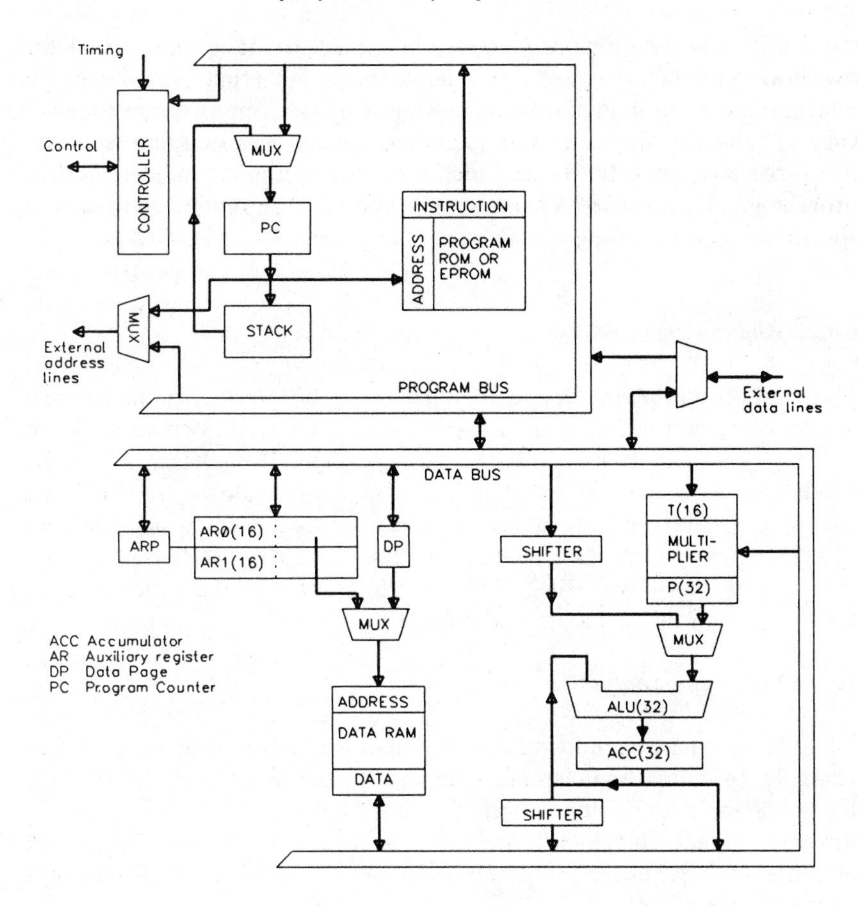

Figure 6.3 DSP architecture

6.1.2 Timer/counters

These are very important devices for a digital controller, because they are usually the primary means of timing, not only for the controller's internal processes, but also for any external event which may be necessary. Their main component is a binary counter, the input to which can be programmed to be derived from either the processor's clock (for timing operations) or from an external source (for counting operations). The counter register can be written to or read from via the processor's data bus, and there may be a number of associated registers for configuring and controlling its operation.

The maximum time period will be determined by the size of the counter register and the frequency of the clock which drives it. Although the counter can be continually read by the processor, it is more usual for overflow of the binary counter to indicate the end of a time period, with an internal flag or an interrupt being used as a signal. Timer/counters have a varying range of programmable facilities; starting and stopping from the processor is normal, but other functions are sometimes provided such as programmable pre-scaler on the clock input and automatic re-load of a specified count following overflow.

6.1.3 Input/output devices

These are the peripheral devices which provide the interface to the physical system being controlled. Some issues have already been briefly mentioned before, but we will now discuss these further. Broadly, they can be subdivided into devices for handling digital or analogue signals. The following sections are mainly orientated towards devices which are connected to a microprocessor bus. In the case of microcontrollers, the input and output 'ports' are such that this 'bus compatibility' may not be necessary, and that sometimes simplifies the design.

(a) Digital input/output

Digital inputs may be individual logic signals for interlocking purposes (for example, from limit switches etc.), or sets of logic signals from transducers having digital outputs (for example, encoders). Digital output is also likely to consist of individual logic signals, such as for enabling or disabling power amplifiers as a part of the interlocking process, or for controlling indicator lights or displays, etc.

The most straightforward digital input device is a set of tristate buffers attached to the data bus and enabled by address decoding responding to a specific address from the processor. The corresponding output device is a set of latches, the inputs of which are connected to the data bus; writing to the latch can be enabled by another specific address from the processor, and clocked in by the processor's Write pulse. These methods are shown in figure 6.4.

The alternative is to use one of a variety of programmable bus-compatible input/output adaptors (often known as PIAs or PIOs); their pins can be individually programmed for input or output by means of additional internal registers. Often they incorporate interrupt handling hardware, and some versions additionally include timer/counters which can help to reduce the hardware complexity of the complete controller. Figure 6.5 shows the overall arrangement of a typical device, with the microprocessor interface to the left and the I/O and control lines to the right.

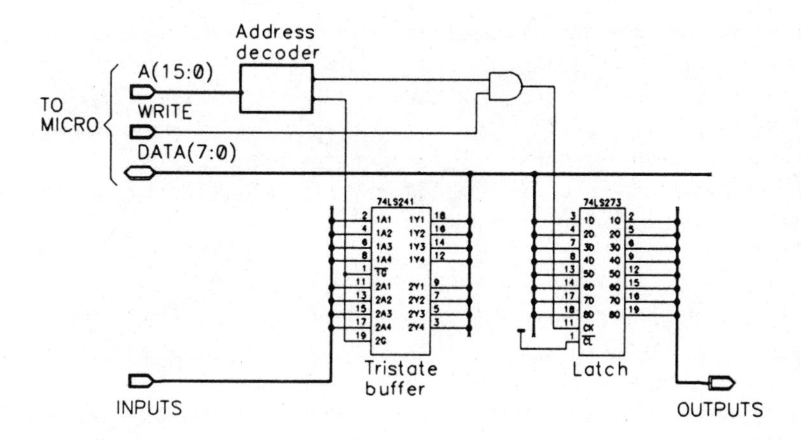

Figure 6.4 Digital I/O circuits

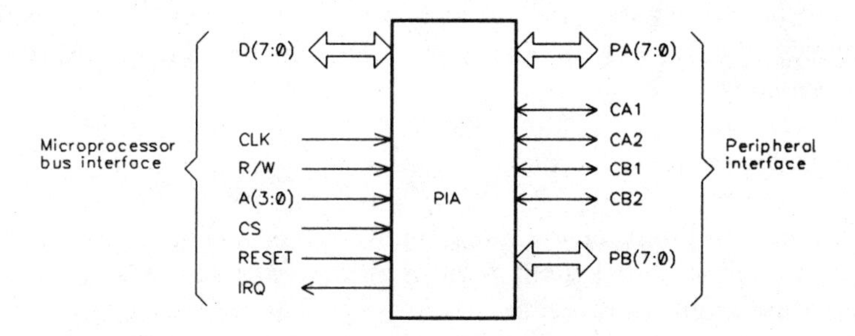

Figure 6.5 Typical programmable I/O device

The signals from absolute encoders can be directly connected by one of the methods mentioned in the previous paragraph, but incremental encoders or tachometers present a different problem because their signals must be counted in order to obtain position information. Bidirectional versions of these produce two digital signals 90° out-of-phase, and theoretically it is simply necessary to decode the direction and count the pulses in order to determine the angle or position. However, in practice, the frequency at which the pulses arrive will be two or three orders of magnitude higher than the controller's sample frequency in order to give sufficient resolution; the software overhead of decoding and counting these pulses will usually be unacceptable. Hence external hardware will be necessary, but bus-compatible tacho ics are available which contain decoding logic and a bidirectional counter, meaning that the processor

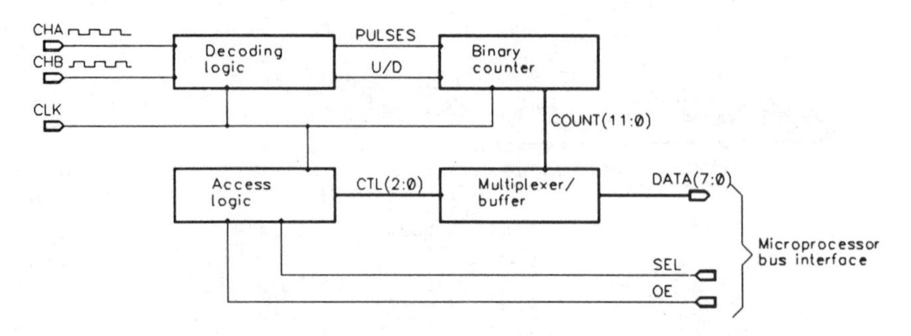

Figure 6.6 Tachometer circuit

merely needs to read the output from the counter at regular intervals in order to extract the angle or position measurement. Figure 6.6 shows the essential arrangement of such a device, the microprocessor input to the right, and the two inputs from the tachometer on the left. Note that the clock which governs the counting process may be derived from the microprocessor's clock signal.

(b) Analogue input and output

The required resolution of the inputs and outputs will have been determined from the overall system performance considerations – anywhere between 8 and 14 bits are possible, with 10–12 bits being most likely. There is a variety of types of A/D converter, giving different trade-offs between the important factors of cost, resolution and speed. Generally speaking, successive approximation converters are most commonly used in control – these, like many of the other peripheral chips for microprocessors, can also be bought for direct interface to 8- or 16-bit data busses. They are a little more complicated than other peripheral devices because additional control is necessary to start the conversion and to indicate that the process is complete (the detail of how they are operated is readily available elsewhere [Co87]). More than one analogue input can be accommodated either by including more than one converter, or by having a single converter with an analogue multiplexer controlled from the processor, or by using a multichannel converter with a built-in multiplexing system. A functional scheme for a typical example of the latter type is shown in figure 6.7.

D-to-A converters are, in general, relatively cheap and simple. Some have in-built output stages, but more usually it is necessary to add an operational amplifer to generate a voltage output of a suitable range. Bus-compatible devices are available which contain an internal latch to hold the signal for the converter itself (figure 6.8). Another possibility is to

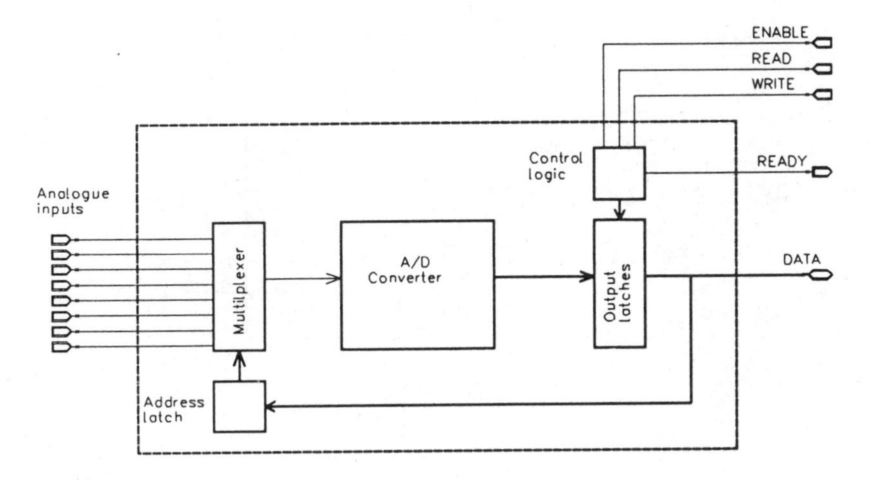

Figure 6.7 Multi-input A/D converter

generate an analogue signal from a single digital output by creating a pulse width modulated (PWM) signal, the pulse width of which is made proportional to the output variable. Taking this signal through a low pass filter generates the analogue output – the overall scheme is shown in figure 6.9. The basic frequency of modulation must be very much higher than the sample frequency if the filter is to reduce the hf component to an acceptable level without causing significant time lags for the control loop. It will clearly take up much of the computational capability to do this in software, but it is mentioned here because some processors designed more specifically for control applications have PWM output stages on the chip. Such output signals can be used directly to control the switching of semiconductor devices in power amplifiers; this is particularly applicable in control systems which incorporate electric motors.

6.1.4 Development and testing

The development stage of a high-performance control system can cover a significant time-span (even though the control circuit itself may have been fully bench-tested), and this fact may have some significant hardware implications. In some circumstances it may be preferable to use a more extensive circuit for the development stage, with extra facilities to ease parameter changes, most likely by means of a terminal connected to the processor. Once proven, these extra facilities can be removed and the circuit design streamlined for production. An alternative would be to use a proprietary development board for the chosen processor.

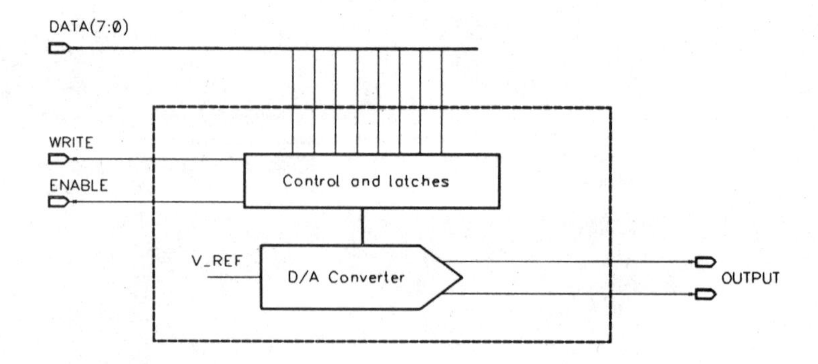

Figure 6.8 Bus-compatible D/A converter

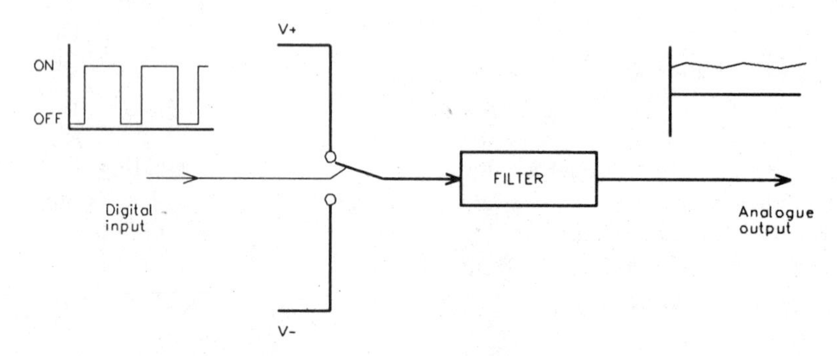

Figure 6.9 PWM output

It is worth emphasising a final point about digital controllers which also may affect the hardware design. In a more complex controller with a number of loops and compensation stages, it is sometimes desirable during testing to monitor internal variables such as control loop error. Whereas with an analogue controller it is usually a simple matter to connect an oscilloscope or meter to a particular operational amplifier's output, the situation is quite different in a digital system. With a normal microprocessor, the internal variables will appear fleetingly on the data bus – in principle it is possible for a logic analyser to capture these, but to reconstruct a time history is very difficult. With a microcontroller they will of course be totally internal and hence inaccessible. If the need for such a monitoring process is expected to be useful, it will be worth building in an extra D/A output channel which may be programmed to access different variables as necessary.

6.1.5 Example control circuits

Three actual digital control circuit designs from the authors' experience are included in order to illustrate some of the broad principles which have been covered in the preceding sections. Consistent with the opening remarks of the chapter, details of each design are not explained, but the overall functioning of the circuitry is. For a detailed understanding it will be necessary to consult data sheets for the devices used, but we believe that the designs will be sufficiently complete to be useful to the reader. The three designs illustrate the use of the three different types of processor which were summarised in section 6.1.1.

(a) A microprocessor design

This example (figure 6.10) has one analogue input, an incremental tachometer input, and two analogue outputs. The processor itself, a Motorola 6809, is one of the more advanced 8-bit devices which includes a multiply instruction and a number of useful instructions for implementing multiple-precision operands. Its data bus is directly connected to five devices: RAM, EPROM, one 10-bit A-to-D converter and two 'VIAs' (Versatile Interface Adaptors), the latter being enhanced peripheral interface ics each containing two 16-bit timer/counters. The circuit was designed prior to the availability of the tachometer processing ics referred to in section 6.1.3, and utilises one counter in each of the VIAs, plus some discrete decoding logic which steers the pulses to one or other counter depending upon the direction of rotation of the tachometer. The digital I/O connections on the VIAs are simply used to connect to low-cost 8-bit D-to-A converters which were perfectly adequate for the application, despite the higher resolution of the input signals. The address decoding makes the digitised input signal, the up and down counts, and the output signals directly accessible to the processor.

(b) A microcontroller design

The second design example (figure 6.11) has a single analogue input and a single analogue output, both with 12-bit resolution. It uses a bottom-of-the-range single-chip microcontroller, the Intel 8748, which includes in-built EPROM. It includes a timer/counter which provides the basic timing functions and is used to determine the sample rate, but otherwise everything is handled in the software. A particular feature of the design is the modular approach which enables further microcontrollers and A/D converters to be added such that multiloop controllers can be implemented [Go89a]; it is equally possible to have a module with a tachometer processing integrated circuit. The design is intended to represent a

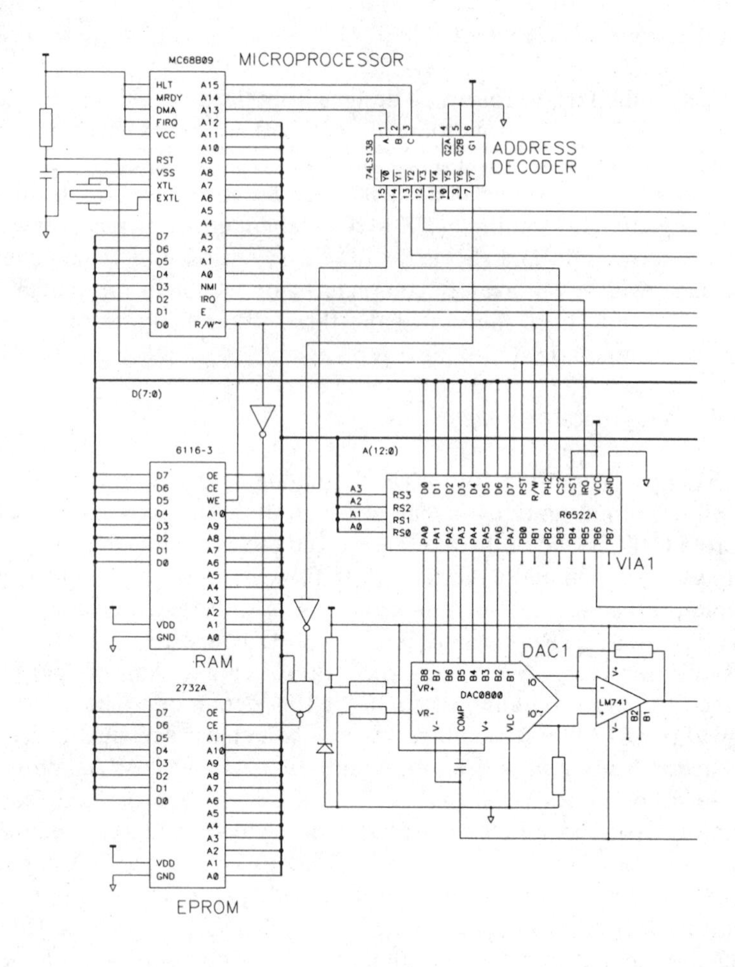

Figure 6.10 Microprocessor control circuit

minimum hardware configuration, but can nevertheless be effectively programmed to give good performance.

(c) A DSP design

The final design example (figure 6.12) uses a first generation Digital Signal Processor, but a variant which provides a number of on-chip features which are targeted for control applications, in particular a latched input/output port and a set of Pulse Width Modulated output drives. The circuit provides eight 12-bit analogue inputs (via an independent multiplexer), a

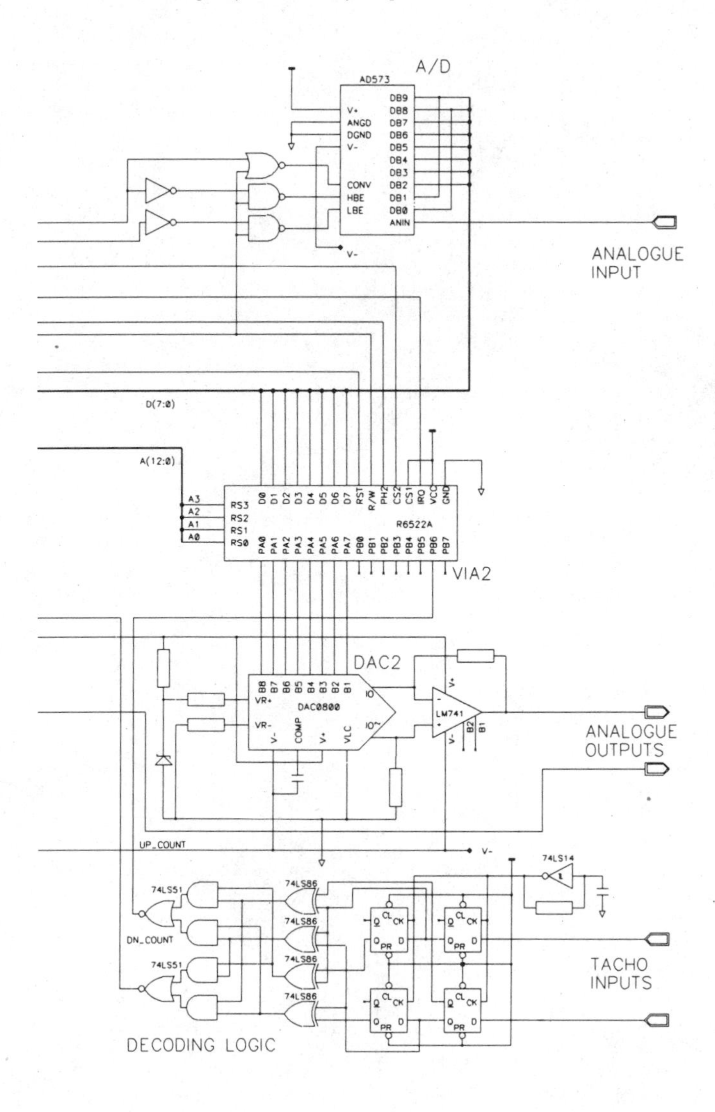

single tachometer input, and two analogue outputs. The latter make use of the PWM outputs, which drive analogue switch circuits to give accurate switched voltage levels to feed into the following filter stage (working directly off the logic output signals from the DSP is unlikely to give high enough quality because of the variability of their voltage levels, although this decision would clearly be application-dependent). There is also provision for a communications link to a terminal using the device's internal serial input/output port.

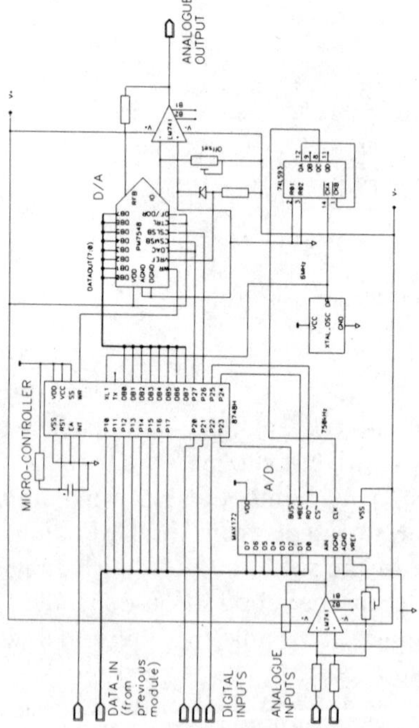

Figure 6.11 Microcontroller control circuit

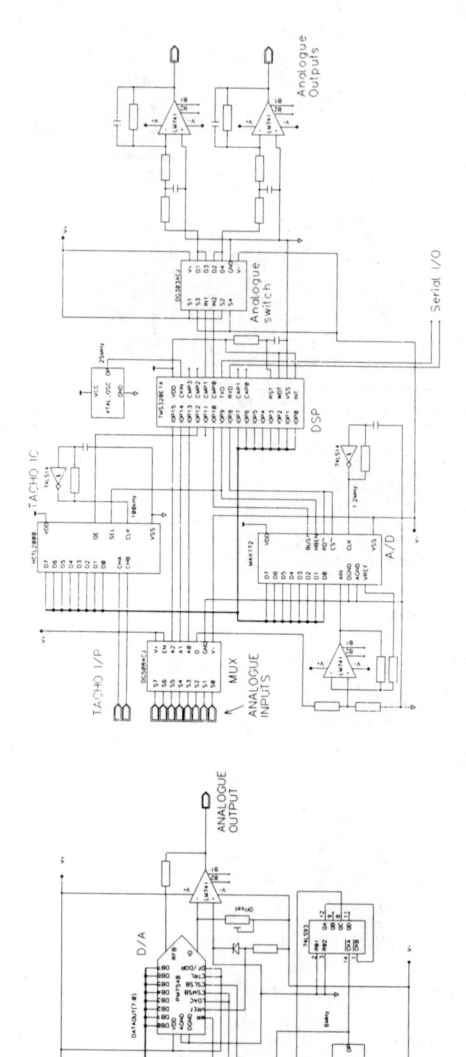

Figure 6.12 DSP control circuit

6.2 Software considerations

The aim here is not to tell people how to design and write software – there are many comprehensive texts which give excellent guidance both in software design methods and in specific languages. However, some of the issues have special significance in the context of digital control, and these can be emphasised while maintaining independence from specific processors or languages. We have included three main topics: firstly an indication of the overall structure of the software, secondly a discussion of numerical routines, and finally a comparative overview of languages for the actual coding of the software. We have tried to achieve a balance which is useful enough to act as a starting point, but without being too detailed, although we have included examples which are (inevitably) specific to a particular processor. We continue to use the C programming language in order to present the organisation of the procedures which must be implemented, although we have restricted ourselves to the most common programming constructs, all of which can readily be translated to a different language, largely by making syntactical changes.

6.2.1 Software structure

The equations which must be executed within each sample period have already been identified, and it is relatively obvious how these need to be coded, depending upon the language used. However we also need an overall structure, a framework upon which to hang these equations, bearing in mind that we may have more than one control loop, and that there may be tasks other than the primary control algorithm.

One of the main objectives must be to achieve a constant sample rate for a given control loop, and all the schemes which follow assume one or more timer/counters running independently of the processor but accessible to it. In principle, it is possible to dispense with the timer and insert dummy operations which 'pad out' the program in order to take the required length of time, but in practice, particularly where there are conditional operations, this may be difficult because the route through the program may vary depending upon a number of factors, including perhaps the current values of the input which cannot be allowed for at the design stage. Such an approach is difficult to get right, yields a program which may be complicated to change, and will probably preclude the idea of carrying out background tasks. We offer a total of five practical software schemes, and discuss their operation, advantages and disadvantages. These will not cater for all applications, but should provide the basis from which the reader's specific requirements can be developed. All the schemes ignore certain functions where they are not pertinent to the scheme's essential operation, such as initialisation of RAM etc.

We have chosen to use the British Standard method of diagrammatic representation, known as Design Structure Diagrams. Appendix H gives a summary of the main features of this method and how they should be interpreted, although the program flow is quite natural and should therefore be fairly obvious [BS87].

(a) Software scheme 1 – main program only

In this scheme (figure 6.13) the first action is to start the timer running (that is, enable a clock input into its binary counter). Note that the size of the counter (8 bit, 12 bit, 16 bit, etc.), and the frequency at which it is being clocked, will have been appropriately chosen at the hardware design stage to suit the required sample period. The program then enters a repetitive loop, the first action of which is to await a 'real-time event', that is, the timer overflowing to indicate the start of a new sample period. This scheme requires the processor to watch the timer continuously, and so it can do nothing else.

Once the timer overflow has been detected, the timer must be re-loaded with whatever value is needed to maintain the correct sample rate. With some timers this re-loading is automatic, the value to be re-loaded having been stored in one of the timer's registers during initialisation.

Once the timer has overflowed and been re-loaded, the equations which implement the control strategy can be computed, including procedures for

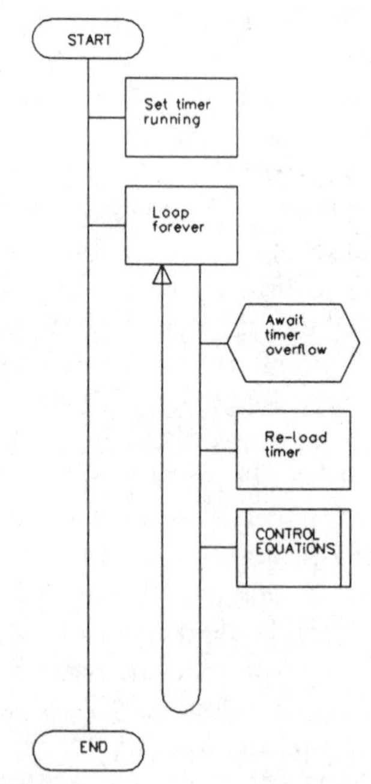

Figure 6.13 Software scheme 1 – main program only

reading the input and writing the output drive signal to the DAC. Once this has all been done, the program loops back to await the next timer overflow.

The C program that follows illustrates how this might be achieved; the procedure control( ) would contain the control equations, I/O operations, etc. (note that in C != means 'not equal to').

```
main()
{      start_timer();
       while (time < end_of_time)
       {    while (timer_overflow() != 1); /*await timer overflow*/
            load_timer(period);
            control();                      /*compute control eqns*/
       }
}
```

Apart from its basic simplicity, the scheme's other advantages are precision of timing (assuming that the equations can be computed within the sample period, of course) and a low overhead for achieving the timing. Its major disadvantage is that not a lot else can be done, apart from quite simple processes which can be packed into the remainder of the sample period.

(b) Software scheme 2 – main program with an additional function

This scheme (figure 6.14) initially operates in a similar manner to scheme 1 in that, having set the timer running, it enters a repetitive loop. However, within this loop the timer overflow is conditionally checked, rather than waited for. If the timer has overflowed then the control equations etc. are computed; if not, the program moves on to another process (shown as an Interlock function, which would be fairly typical), which is repetitively executed until the timer has overflowed. A corresponding C program is:

```
main()
{     start_timer();
      while (time < end_of_time)
      {     if timer_overflow()
            {     load_timer(period);
                  control();
            }
            interlock();
      }
}
```

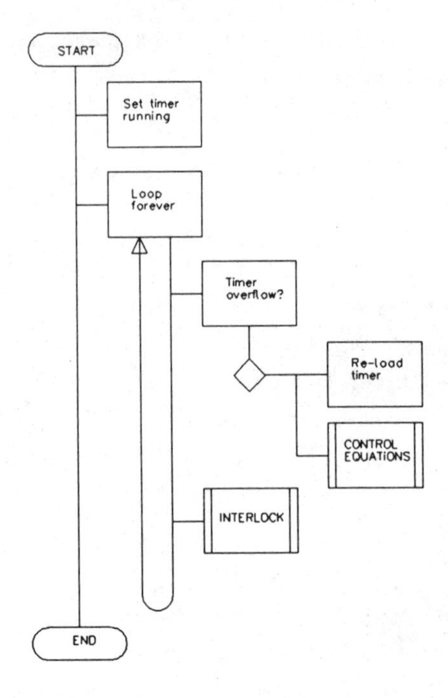

Figure 6.14 Software scheme 2 – main program with other function

The advantage of this scheme is that it enables another process to be carried out within the time spent waiting for the timer to overflow. Its disadvantage is that a length of time elapses after the timer overflows before the control equation is executed, a period of time that depends upon how long the other process takes. If the additional process takes a significant proportion of the sample period, then this method is unlikely to be suitable. If it takes longer than the sample period, then it is certainly unsuitable.

(c) Software scheme 3 – main program plus timer interrupt

In the previous two schemes, the processor would need to be explicitly programmed to check for the timer overflowing. The alternative (figure 6.15) is for the timer to generate an interrupt when it overflows such that the processor stops what it is doing, stores away whatever data it was working on and jumps to a totally different piece of program code. On completing this 'interrupt routine', it retrieves the relevant data and carries on where it left off.

The main program first activates the timer interrupt, shown as a 'task', and then moves to an endless loop in which other processes may be carried

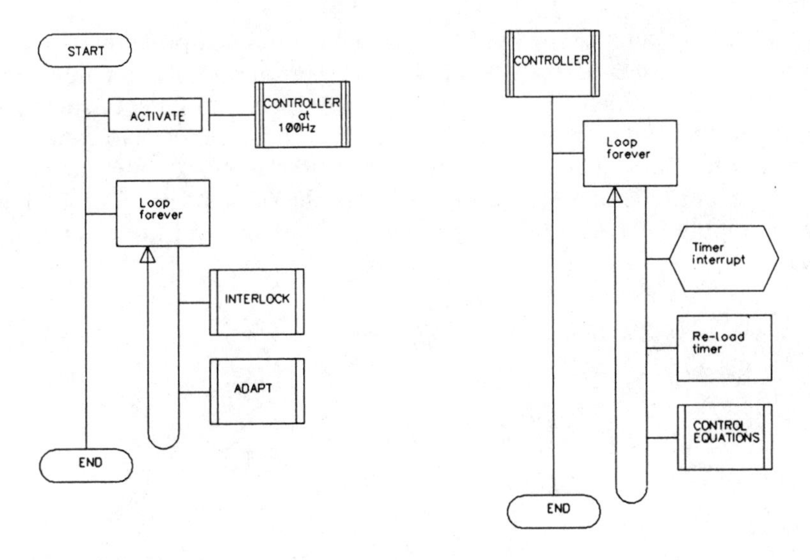

Figure 6.15 Software scheme 3 – main program with timer interrupt

out independently of the computation of the control equations – processes 'interlock' and 'adapt' are shown as typical examples. Note that these may take significantly longer than the sample period, and so either or both may be interrupted a number of times in the course of their execution in order that the control equations may be computed at each sample instant.

A separate diagram shows the Controller task. Note that it is not a straightforward matter to show interrupts in software diagrams. The method used is that recommended by the authors of the British Standard referred to previously, by which the interrupt is shown as a real-time event within a repetitive loop; the control equations are executed after the interrupt arrives. The code which comprises the interrupt routine will simply end in a 'Return from Interrupt' instruction, and therefore will not explicitly include the repetitive loop shown in the diagram, which at first sight suggests that the representation may be wrong. However a processor's ability to respond to an interrupt is usually contained within the 'microcode' which is executed every time an instruction is decoded, and therefore the repetitive loop is effectively in the software. It is not specifically incorporated by the software designer, simply because it is inherent within the processor.

Handling of real-time processes such as interrupts is essentially a 'low-level' activity, possibly best programmed in Assembly language, although constructs for interrupt handling from high-level languages are sometimes available. The following sequence comprises a mixture of Assembly language and C instructions. 'Reset_vector_address' and

'Timer_vector_address' are pre-defined entry points in program memory to which the execution will jump after power-on and a timer interrupt respectively. Normally these entry points are used with jump instructions (JMP) to the actual program address from which execution is to continue, and are often known as 'vector addresses'. 'ORG' is an Assembly language directive which defines the starting point for the subsequent code. 'RTI' is an Assembly language instruction which signals the end of an interrupt routine.

```
        ORG     Reset_vector_address
        JMP     Start

        ORG     Timer_ vector_address
        JMP     Timint

Start:  activate();          /*Initialise timer interrupt etc*/
        while (time < end_of_time)
        {    interlock();
             adapt();
        }

Timint: load_timer(period);
        control();
        RTI
```

The mix of Assembler and C shown above would not work because extra compiler directives would normally be needed to switch between the two types of instruction. Including such directives would however create confusion, and in any case are compiler-specific, and so they have been deliberately excluded in order to illustrate the diagrammatic scheme of figure 6.15 in a pseudo-code form.

The advantage of using a timer interrupt in this manner is that it creates independence between the primary control task and other background processes, and frees the designer from having to sort out the relationships which would otherwise be necessary; however if there is the need to transmit information between the two, considerable care must be exercised. The main disadvantage is the overhead associated with the handling of the interrupt, predominantly that of saving the exit point from the program being executed when the interrupt arrives; in addition there is the need to store the accumulator, internal registers and anything else which might be corrupted within the interrupt routine. Also, of course, there is the complementary process of retrieving the data and program address once the routine has been completed.

The preceding three schemes have been concerned with single control loops, but it is essential that methods for multiloop controllers using a single processor are also dealt with. The following two schemes apply to the two loop control system shown by figure 6.16, and can be readily extended to systems having a larger number of control loops, and to systems having other than the 'loops within loops' arrangement. (The assumption here is that the loops will have different sample rates, which may often be the case. If they have the same sample rate, one of the previous schemes is suitable.)

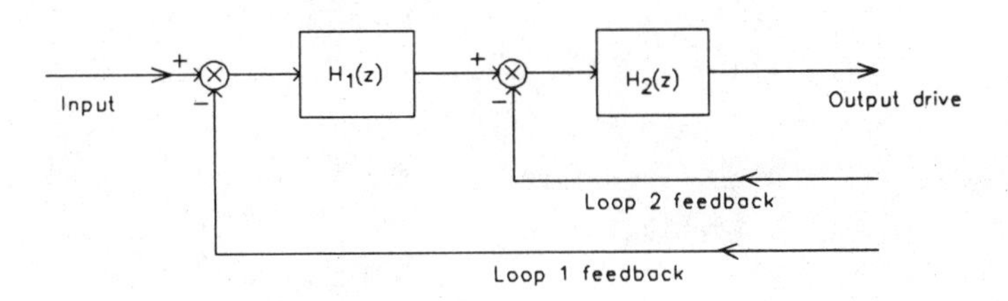

Figure 6.16 Two loop control system

(d) *Software scheme 4 – Main program with two independent timer interrupts*

This is a relatively straightforward adaptation of scheme 3, in which the possibility of multiple interrupt sources is allowed for within the real-time event (figure 6.17). Discrimination between the interrupt sources will be inherent within the processor in the case where different interrupt vector addresses are provided, but must be programmed where only a single vector address is available. The former is often the case for a microcontroller in which the timer–counters are within the integrated circuit, whereas for a microprocessor which has two external timer/counters activating a single interrupt request line it will be necessary to identify the interrupt source. The two pseudo-programs which follow indicate how this might be done:

Separate vector addresses:

```
ORG     Reset_vector_address
JMP     Start

ORG     Timer1_vector_address
JMP     Timint1
```

```
                ORG         Timer2_vector_address
                JMP         Timint2

Start:          activate();
                while       (time < end_of_time)
                {           interlock();
                            adapt();
                }

Timint1:        load_timer1(period1);
                control1();
                RTI

Timint2:        load_timer2(period2);
                control2();
                RTI
```

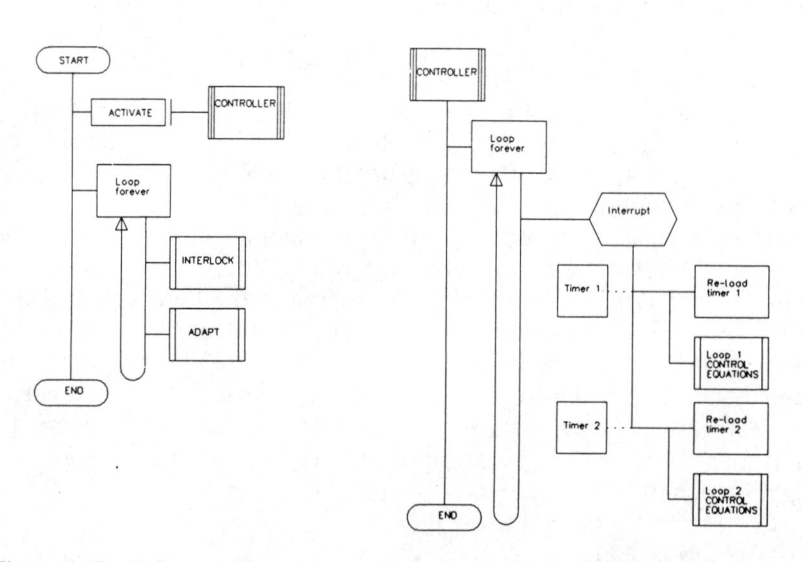

Figure 6.17 Software scheme 4 – main program with two timer interrupts

The output of control loop 1 will be used as an input to control loop 2. Since the timer interrupts are asynchronous, this transfer of data must be thoughtfully handled. It is likely that one interrupt routine would be completed even if the other interrupt occurred during its execution, as

Single vector address:

```
                ORG     Reset_vector
                JMP     Start

                ORG     Timer_vector
                JMP     Timint

Start:          activate();
                while   (time < end_of_time)
                {       interlock();
                        adapt();
                }

Timint:         if      timer1_interrupt
                {       load_timer1(period1);
                        control1();
                }
                if      timer2_interrupt
                {       load_timer2(period2);
                        control2();
                }
                RTI
```

shown by figure 6.18(a), in which case the use of a common or 'global' variable, stored by the loop 1 routine and loaded by the other, would be sufficient. Sometimes it may be desirable to give a higher priority to the faster control loop (usually the inner one) in order to ensure that the computations are carried out with the minimum delay; in this case it must be able to interrupt not only the main (background) task, but also the other timer interrupt routine which may be executing the equations for the slower control loop – this is illustrated by figure 6.18(b). In this situation there is an increased danger of corrupting the data to be transferred between the routines, because the slower loop may be in the process of updating as the faster interrupt arrives, and so great care must be exercised to avoid this.

Although not included as a separate scheme, the control loops could be implemented as fully independent tasks which are executed concurrently, rather than as a single task carrying out two functions. (An appropriate multitasking operating system, or at least the procedure which schedules

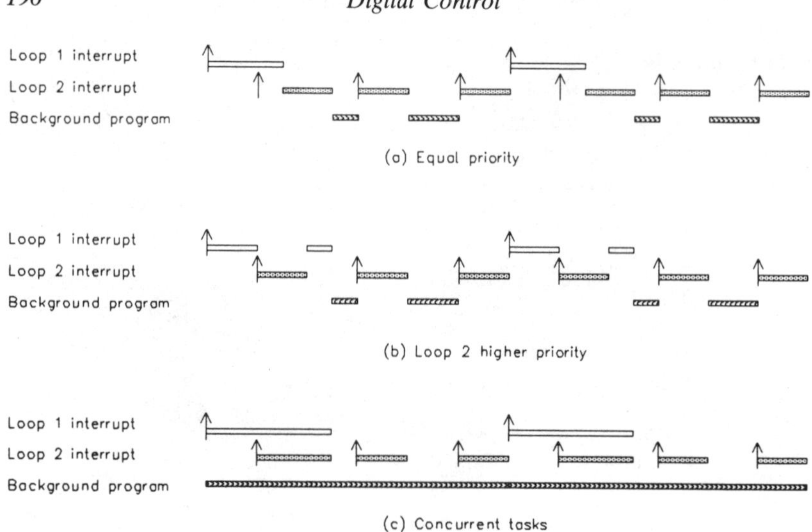

Figure 6.18 Operation of interrupt routines

the tasks, would be needed for this approach, however.) Figure 6.18(c) shows the difference compared with the previous two possibilities – the execution continues without interruption even when the two loops are being computed together, although obviously their execution times will be extended. Transmission of information is further complicated by this method of operation, but may well be provided for by a utility within the multitasking operating system.

This approach using two timer interrupts has the advantage of flexibility, giving completely independent timings for the two loops (or more if the scheme is extended). Remember, however, that the overhead associated with handling the interrupts will be further increased, so it will be necessary to keep a check on what proportion of the total processor capacity is being taken up – too high a proportion will significantly reduce the time available for computation, and may compromise the ability to achieve the desired sample rates.

(e) Software scheme 5 – main program with a single timer interrupt handling the two loops

This scheme (figure 6.19) requires that the slower loop has a sample frequency which is a submultiple of that of the faster loop. This means that a single timer/counter generating interrupts at the higher sample frequency is sufficient. The control equations for both loops are then included within the single interrupt service routine, which must incorporate additional code such that the slower loop is not executed every time. As with the

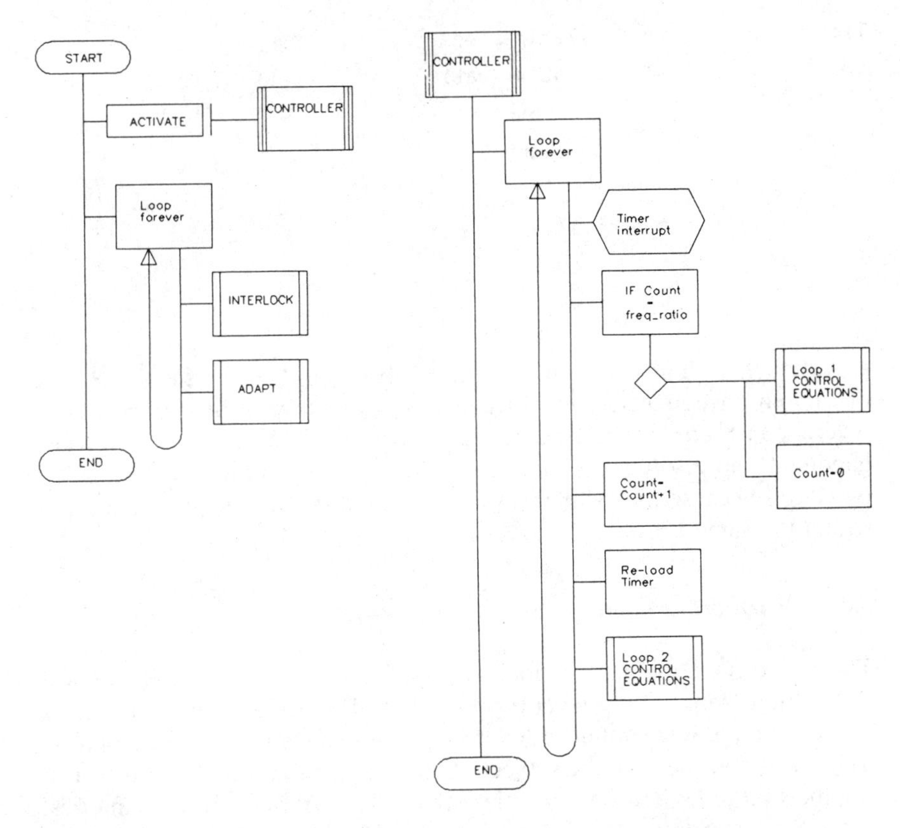

Figure 6.19 Software scheme 5 – main program with timer interrupt (two loops)

previous schemes we are including a suggested program, again a simplified mix of Assembler and C code to illustrate the principles:

```
                ORG      Reset_vector_address
                JMP      Start

                ORG      Timer_vector_address
                JMP      Timint
Start:          activate();
                while    (time < end_of_time)
                {        interlock();
                         adapt();
                }
```

```
Timint:         load_timer(period);
                if  (count == ratio)        /* ratio = f2/f1*/
                {       control1();
                        count=0;
                }
                control2();
                count=count+1;
                RTI
```

Although such a scheme is less flexible because it constrains the choice of sampling frequencies, in practice this is unlikely to be a particularly severe constraint. It offers a reduced overhead in terms of interrupt handling, and also considerably simplifies the data transfer between the two loops because they will always be executed in a defined sequence and within the same routine.

6.2.2 Numerical routines

The previous section has provided an appreciation of how the software will fit together from the top-level point of view. The actual control equations (see Chapter 5) were not included because these depend upon the language to be used, but they are the subject of the next section. Firstly, however, it is important to review the principles involved in carrying out the numerical routines on a processor; we then relate these to the conclusions of Chapter 5 in which the precision of computation was determined.

Once again, an in-depth coverage of the topic is not intended, but we hope to promote a degree of understanding which can be reinforced where necessary by texts giving more working detail on binary arithmetic in processor systems. We briefly discuss number representations, and then move on to the principles of addition, subtraction and multiplication, particularly with respect to multiple precision operations.

As observed in section 5.10, the wordlengths for the coefficients and the internal variables have been determined irrrespective of the specific processor to be used. These wordlengths represent minima of course, because we will be using multiples of 8 or 16 bits (and sometimes perhaps 32) – consequently we will often get more precision than is strictly needed, but nevertheless we are assured of the digital filters' proper operation.

The kinds of calculation which have been identified require the addition and subtraction of fixed-point variables, and also the multiplication of a fixed-point variable by a coefficient. The coefficients are sometimes well represented in fixed-point format but, particularly for the δ-filters, the potential for using a simple floating-point coefficient has been highlighted.

Basic digital control using time-invariant discrete transfer functions can be most effectively achieved using only these three routines. In more sophisticated controllers, for example those involving adaptation, there may be more complex requirements such as the multiplication of two variables, but these are beyond the scope of this text. The following review is written with these basic needs in mind.

(a) Number representations

Unsigned integer numbers are represented in natural binary, and a wordlength of i bits will give a number n with a range $0 \leqslant n \leqslant 2^i - 1$. For example:

$$i = 8 \qquad 0 \leqslant n \leqslant 255$$
$$i = 16 \qquad 0 \leqslant n \leqslant 65535$$

Signed integer numbers can be represented by a natural binary number plus an extra sign bit (usually 0 for positive, 1 for negative), but it is more usual to employ the two's complement representation in which positive numbers occupy the bottom half of the range, and negative numbers the top half. A wordlength of i bits will give a number n with a range

$$-2^{(i-1)} \leqslant n \leqslant +2^{(i-1)} - 1$$

For example

$$i = 8 \qquad -128 \leqslant n \leqslant +127$$

$$i = 16 \qquad -32768 \leqslant n \leqslant +32767$$

Positive numbers are directly represented; negative numbers are represented by adding 2^i, which is equivalent to complementing every bit and adding 1 lsb to the result. Hence, for $i = 8$:

$$0 = 0000\ 0000_B = 00_H$$

$$127 = 0111\ 1111_B = 7F_H$$

$$-1 = 1111\ 1111_B = FF_H$$

$$-128 = 1000\ 0000_B = 80_H$$

The most significant bit indicates the sign: 0 for positive, 1 for negative.

Note that A/D conversion of bipolar analogue signals will often give what is called 'offset binary' representation, in which the most negative is converted to all 0s, and the most positive to all 1s. It is easy to see, however, that simply complementing the msb will convert to 2's complement representation.

Unsigned and signed non-integer numbers can readily be adapted from the above by positioning the binary point other than at the right-hand end of the number. As long as its position remains fixed, it can easily be allowed for in the arithmetical operations.

Floating-point numbers consist of a mantissa (which may be signed or unsigned) and an exponent (which will nearly always be signed so that numbers both greater than and less than unity can be accommodated). The wordlength of the mantissa determines the precision to which the number can be represented, and that of the exponent determines its dynamic range; it is not necessary for the mantissa and exponent to have the same wordlengths. There is an IEEE standard which defines the formats for binary floating-point numbers which are often used by high-level languages, although clearly low-level Assembler programming is not limited to these.

Previous considerations relating to overflow of the internal variables have neglected the idea of negative numbers, but given a certain wordlength for the input and output variables, it does not in fact change anything. The maximum size of the input is halved (for example, −2048 to +2047 instead of 0 to 4095 for 12-bit I/O resolution), which halves the size of the overflow; however, a bit is still needed for the sign of the internal variable, and so the overall size is unchanged. In general, the handling of negative numbers tends to complicate the issues a little, but does not fundamentally change them.

(b) Addition and subtraction

Normal binary addition (and subtraction) is the same for both unsigned integers and signed 2's complement integers; the difference between the two types lies in the way overflow is handled. Addition of two unsigned integers may generate a 'carry', and subtraction may require a 'borrow' – in processors these are both usually indicated by the same internal flag, and any corrective action which needs to be applied will depend upon the operation which caused the carry/borrow. Addition and subtraction of two signed integers may also set the carry/borrow flag, but the decision as to whether this is an overflow depends upon the signs of the two original numbers. For example, on an 8-bit processor the addition of – 1 and – 2 will set the carry/borrow flag (on an 8-bit processor these will be represented as FF_H and FE_H respectively), and yet clearly the result does not overflow the maximum range. Some processors have a 2's complement overflow flag which simplifies things.

Chapter 5 has shown how to determine wordlengths for the internal variables, and from the examples which were presented it is clear that multiple-precision operations will often be necessary for 8- and 16-bit processors, although obviously this will depend upon the input/output wordlength and the sampling frequency. Multibyte or multiword additions/subtractions are relatively straightforward for both signed and unsigned variables; it is necessary to add/subtract the least significant byte or word first, and then continue up to the most significant, including the carry or borrow from the previous operation. Figure 6.20 illustrates the process for a triple precision operation. Overflow only needs to be checked for (and, if necessary, a limit applied) after the last addition has been done.

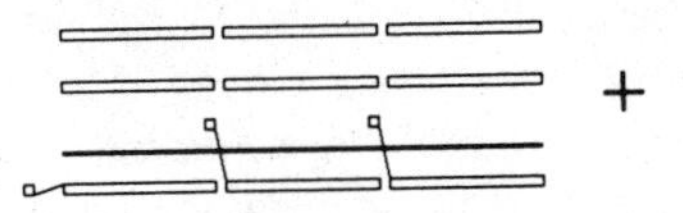

Figure 6.20 Multibyte addition

The above considerations apply to the addition and subtraction of fixed-point variables having a common position for the binary point. Different positions can be dealt with by left or right shifting one or other of the variables so the points line up prior to the operation. Addition and subtraction of floating-point variables also requires this alignment process: before the operation takes place, the exponent of the smaller number must be adjusted to equal that of the larger. As a result, the mantissa of the smaller number will contain leading zeros and will thus have lost a degree of precision. For this reason, floating-point arithmetic may not yield the precision expected – what is gained in the multiplications may be lost in the additions. The true precision of a floating-point multiply is what is left after any following addition.

(c) Multiplication

The starting point will normally be the multiplication of two bytes or words, depending upon processor type. If the processor has a multiply instruction the product will be double length, as shown by figure 6.21(a). If it does not have a multiply then a routine must be programmed, normally using a 'shift-and-add' procedure. The multiplications may be signed or unsigned, and these are distinct operations. If non-integers are being

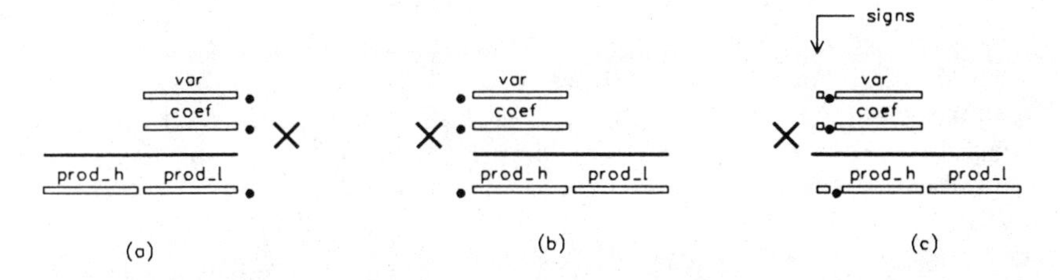

Figure 6.21 Single-precision multiplication

multiplied, it is necessary to make an appropriate adjustment in the position of the binary point in the product, and figure 6.21(b) shows the result with purely fractional *unsigned* numbers. Notice however that the binary point in purely fractional *signed* numbers is in effect between the top two bits, which means that the result of a signed multiplication operation is different, and is shown in figure 6.21(c).

For digital controllers, although the variables will almost certainly change sign, the sign of the coefficients is pre-determined and can therefore be accommodated by programming either an addition or a subtraction of the product of the multiplication, and consequently only positive coefficients strictly need to be allowed for. If an unsigned multiplication is being used (see figure 6.21(a)), it can easily be shown that the result should be corrected by subtracting 'coef' from 'prod_h' if 'var' was negative.

Multiple-precision multiplications need some mention, because it is quite likely that these will be necessary, even with a 16-bit processor. Double precision on an 8-bit processor would imply the multiplication of a pair of 2-byte variables, yielding a 32-bit result, and figure 6.22 demonstrates how this would be formed. It can be seen that this process requires four multiplication operations and five additions (two additions to form 'result_lh', two to form 'result_hl', and a final addition to include the carry into 'result_hh'). The reader will find that the effect of a negative variable can be corrected by subtracting the coefficient from the top two bytes of the result.

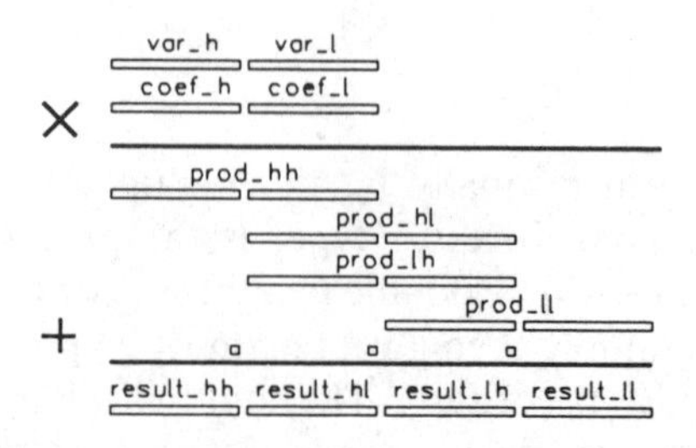

Figue 6.22 Double-precision multiplication

The multiplication used four times within this operation is assumed to be unsigned; a signed multiplication, if used in this situation, will obviously cause difficulties because the most significant bit is only a sign bit in *some* of the words. Multiplication of floating-point numbers requires the mantissae to be multiplied and the exponents to be added – note, though, that the product of the mantissae will be rounded to maintain the same wordlength, and hence to keep the same precision.

Having discussed the general issues relating to multiplication of binary numbers, it is useful to become a little more specific to the computational needs which the methods of the previous chapter have determined. Although it is normal that the basic multiply instruction in a general-purpose microprocessor will multiply equal length numbers, and that standard multiple-precision routines similarly apply to equal-length numbers, there is no particular reason that this should be the case in all computations. What we have found is that the internal variables need to have significantly longer wordlengths than the coefficients – this is particularly the case with the δ-filter.

The underflow calculations carried out in Chapter 5 indicated that the new value of an internal variable formed by adding a number of products should be truncated (or rounded) to the same length as the original variable. In Chapter 4 it was pointed out that errors are minimised by performing the truncation *after* the summation, rather than directly upon obtaining the individual products, and this was the basis assumed in Chapter 5. In practice, it is found that truncation *before* summation is quite a lot simpler and that provided the underflow problem has otherwise been properly addressed, as explained in Chapter 5, the question of truncation before or after summation loses its importance. Hence, we might for example require a triple-length variable multiplied by a double-length coefficient; full precision will be a quintuple-length result, and figure 6.23 illustrates the way this would be carried out, including how a triple-length variable could be taken from the full result. Notice that the operations falling to the right of the dotted line may be neglected, although there may then be some small inaccuracy in 'result_l' due to the lack of carries from the lower order addition. This 3 × 2 multiplication can be seen to need five multiplies and six additions (three to form 'result_l', two to form 'result_m'

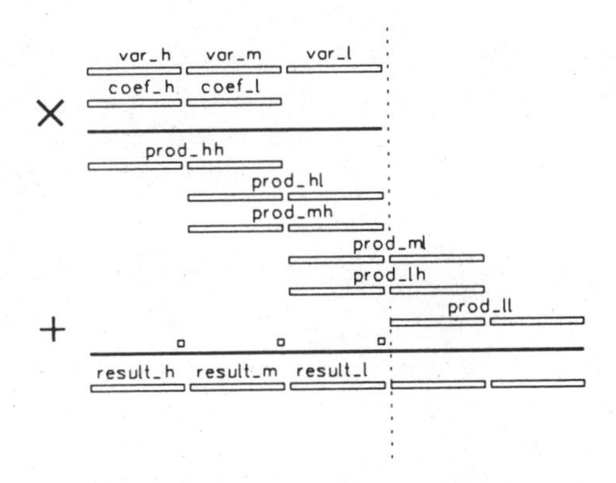

Figure 6.23 3 × 2 multiplication

and one to include the carry into 'result_h'). An additional multiplication and addition would be needed to obtain the full product.

Remember that so far we have made no assumptions about the programming language; it may be that the control system designer will not write the numerical routines, either because a high-level language is being used or because a software specialist is doing the programming. However it is essential that the arithmetical processes which will be invoked are understood, whether they are implicitly included in the software, or explicitly programmed in.

6.2.3 Languages

Fundamentally, the choice is between a high-level compiled language with instructions and syntax which are independent of the processor, and a low-level Assembly language using an instruction set which is specific to the processor. The preference will depend upon a number of factors, including the sample rate, the complexity of the control strategies, the efficiency of the compiler for the high-level language, and the versatility of the processor's basic instruction set. The choice is not clear cut, because high-level programming can usually include low-level instructions, while there are constructs within good Assemblers which provide higher-level facilities.

In the following three sub-sections, the advantages and disadvantages of the choices are summarised and contrasted. Also possibilities within each which can mitigate some of the disadvantages are identified. Some examples from our own experience are presented.

High-level compiled languages

The advantage of a high-level approach is that the program code is compact, easy to read and generally quick to develop. Having written and tested a program it can in principle be used with any other processor for which a compiler is available, that is to say, it is 'portable'. Portability is not however a straightforward matter, and is an important issue among high-level languages.

One disadvantage of high-level languages is that a digital controller will inevitably incorporate peripheral devices, in particular ADCs and DACs; high-level instructions are not generally well suited to handling these, although the versatility of modern languages is improving with provision for real-time operation. However the major disadvantage of using a high-level language is inefficiency, mainly in terms of the utilisation of the processor's capability, because this inefficiency will always result in increased computation time, often very significantly. In nearly all cases, the program memory (EPROM) requirements will be increased, and often extra data memory (RAM) will be needed. Whether these inefficiencies matter in a project sense depends entirely upon the application, a point which is taken up again later.

It is worth exploring the reason for the inefficiency of a high-level language, mainly because the implications will help in the choice of data types for high-level implementation. In a high-level language, one is restricted to certain data types which may be either integer or floating-point numbers, and several possibilities exist within these two categories. For example, C and Pascal offer 8-, 16- and 32-bit variables, both unsigned and signed. Floating-point numbers in the two languages are not exactly the same: in C the simplest is the 'float' data type which has a sign bit, a 23-bit mantissa and an 8-bit exponent, whereas in Pascal the simplest is a 'real' which has a mantissa of 39 bits. There is also a 'double' type in both languages with a sign bit, a 52-bit mantissa and an 11-bit exponent. Other data types giving a much larger exponent are also generally available, but these are unlikely to be wanted in digital control applications.

It is necessary to choose from these data types when assigning variable names and using them in equations, but it must be remembered that the more complex higher-precision data types increase the inefficiency to which we have referred. Basic arithmetic operations such as addition, subtraction and multiplication require that both operands are of the same type (if not, the compiler inserts extra code to convert the lower precision operand to the same data type as the higher). Furthermore, the *result* of the particular operation will be of the same data type, and it is the responsibility of the programmer to choose a data type which will avoid overflow. There may in some cases be the option to build-in error-checking code, but this will not prevent a failure in the computation. The implications of these basic rules can be highlighted by two examples of the variable/coefficient combinations which were identified in the previous chapter.

A brief recapitulation on the conclusions of Chapter 5 is useful before considering the two examples. Wordlengths were identified for the coefficients and the variables, the former to satisfy the requirements for accuracy, and the latter to ensure proper operation with both the largest and smallest changes at the input. As was pointed out, the variables are repetitively multiplied by the coefficients in the recursive operation of the discrete filters; although there is a definite limit to the overflow, the successive multiplications can create an ever-increasing number of underflow bits, and truncation is inevitable. The analysis of Chapter 5 defined the level at which this truncation should occur. Consequently the multiplication process only needs to be carried out to the precision of the original variable, and the position of the product's binary point should coincide with that of the variable.

In example 5.1, the phase advance filter using the z-operator needs a 21-bit $\times$ 11-bit multiplication to form the internal variable. If we first consider using integer arithmetic, the length of the variable requries a 32-bit data type ('long integer' in Pascal, 'long' in C), and it must be signed so that negative numbers are catered for. The coefficient should also be of the same type; if it is not, then the compiler will build-in a conversion procedure which will involve extra computation time and program code. For the time being we are forgetting the positions of the binary point, and just dealing with integers, so the variable, which should have 5 fractional bits and 16 integer bits, must be pre-scaled by 2^5; the coefficient, which should have 8 fractional bits and 3 integer bits, will be pre-scaled by 2^8. Having 21 significant bits in the variable and 11 in the coefficient, it is obvious that the result cannot overflow the 32-bit data type. It will then be necessary to divide the result by 2^8, or right shift 8 times, so that the position of the variable's binary point is preserved. Some of the low bits are lost in this process, but the effect has been allowed for in the underflow calculations. Note that a general-purpose 32 $\times$ 32-bit procedure has carried out the multiplication. However, with a 16-bit processor, a 32 $\times$ 16-bit multiplication would have been sufficient, needing only 2 multiplications instead of 4; for an 8-bit processor, a 24 $\times$ 16-bit multiplication is all that would have been necessary, giving a reduction from 16 down to 6 multiplications.

For completeness, a control procedure in C is given in table 6.1 which could implement a single loop controller with the phase advance compensator.

The following comments clarify what is being done:

(a) The use of 'static' data types for the coefficients preserves their existence between successive executions of the control equations. This avoids the time overhead of creating them each time the procedure runs.

(b) The use of a 'static' data type for variable v1 (initialised to zero) is essential, otherwise the stored value may be over-written before it is used at the next sample instant. This is not necessary for the other variables, neither do they need to be initialised.

(c) The inputs are pre-scaled by 2^5 to allow space for the underflow bits. This is carried out by 5 left shifts of the values which are returned by the ADC routines.

(d) The results of the multiplications are post-scaled by 2^{-8} to compensate for the coefficient prescaling.

(e) The output must be post-scaled by 2^{-5}, which could have been done as it was calculated. Strictly it should have been rounded by adding 2^4 first – this has been omitted for clarity. The greater than unity high-frequency gain means that an overflow limit should be included, also omitted.

Table 6.1 Single loop control procedure

```
control()
{           /* 32bit coefficients, prescaled by 2^8 */
        static long a0=1231   ,a1= -1207 ,b1= -232   ;

            /* 32bit Variables */
        static long v1= 0 ;
        long input,output,feedback,error,v0;

            /* Inputs, prescaled by 2^5 */
        input = read_adc(1) << 5;
        feedback = read_adc(2) << 5;
        error = input-feedback;

            /* Control equations, including postscaling by 2^-8 */
        v0 = error - (b1*v1) >> 8;
        output = (a0*v0+a1*v1) >> 8;

            /* Output, postscaled by 2^-5 */
        out_dac (output >> 5);

            /* Shift operation */
        v1=v0;
}
```

A second example is Example 5.2, the notch filter using the z-operator, which needs a 28 × 14-bit multiplication yielding a 28-bit result. Although 32 bits will be sufficient for the variable, by the time it has been multiplied by the coefficient the result will have well and truly overflowed, meaning that the 32-bit data type is not sufficient. This forces the use of floating-point variables since these are the only higher precision data types which are generally available, although care is still needed. For example Pascal's 'real' data type, which has a 39-bit mantissa plus sign with an 8-bit exponent, will be suitable; however the C 'float' data type, which has a 23-bit mantissa plus sign, is not precise enough. The use of floating-point numbers avoids the pre-scaling and post-scaling which is necessary when using integer variables, but this must be at a significant cost in terms of the computation time and program code needed to carry out the 40 × 40 multiplication (or higher), even without the extra complexity when the floating-point numbers have to be added. There are some hardware implications here too, because it is sometimes possible to obtain maths co-processors to run alongside the basic processor, which then 'farms out' these more complex calculations, essentially to be carried out in hardware rather than software. Such devices are not cheap and are only available for the more sophisticated processors, but nevertheless should be borne in mind if an application turns out to be computationally intensive.

There are two approaches which may be used to alleviate the inefficiencies which we have just identified. Both involve constructing a more efficient *asymmetric* multiplication, one by means of a purely high-level procedure, the other by writing an Assembly language routine within the high-level code. Take, for instance, Example 5.2 which includes the 28 × 14-bit multiplication: the purely high-level approach would use a procedure with two parameters, both declared as 32-bit data types for the variable and the coefficient, but returning a 32-bit value as indicated in table 6.2.

Table 6.2 C procedure to give 32 × 16-bit multiplication

```
long mult(unsigned long var, unsigned long coeff)
{
      unsigned long var_h, var_l, result;
      var_l = var & 0x0000ffff; /* Remove top 16 bits with Hex mask */
      var_h = var >> 16;
      var_l = var_l * coeff;
      var_h = var_h * coeff;
      if (var >= 0x80000000)    /* Test for var being negative */
            var_h = var_h - (coeff << 16);
      result = (var_h << 3) + (var_l >> 13);
      return (result);
}
```

The procedure must separate the upper and lower 16-bit portions of the variable into two distinct 32-bit variables (the upper half of each consisting of zeros) and multiply each one separately by the coefficient; neither multiplication will now cause overflow. The result must then be scaled to allow for the 13 fractional bits of the coefficient and recombined into a single 32-bit variable. Such a procedure is not straightforward, and care is needed when dealing with the signing of the various results, but it at least preserves the advantages of high-level coding: the overall saving compared with using the floating-point data types would need careful evaluation, however. The alternative is to include a low-level procedure which would have parameters declared as 32 bit and 16 bit for the variable and coefficient respectively, and would return a 32-bit value. Assuming that the compiler can cope with Assembler routines, the interface between the high and low levels, usually via the stack and/or some internal registers in the processor, should be well described in the appropriate manuals. The Assembly language instructions will however be processor-specific, which detracts from the advantage of an otherwise high-level approach, but will certainly give a faster execution time.

The preceding comments have been written on the assumption of either procedural or functional constructs in the language, which is usual, but the reader should be aware that there are other languages such as Forth which are lower level and hence do not necessarily suffer from the inefficiencies which we have described; however it is not possible to write mathematical equations in Forth in the normal manner, which means that the ability to read Forth code is a specialist skill and hence is inappropriate for this book.

Low-level Assembly language

The advantage of using Assembly language is that the code is written directly in the instructions which the processor executes. There is no necessity to use specific data types, and it is possible to streamline the arithmetical routines exactly according to requirements, which gives the highest possible efficiency both in execution speed and memory requirements, subject of course to the manner in which the software is designed and written. Dealing with peripheral devices such as A/D and D/A converters is usually very straightforward in Assembly language.

There are however disadvantages to low-level programming: fundamentally the lack of portability, and the likelihood of increased development time involved in writing the arithmetical routines. Remember, however, that for basic digital control strategies such as are covered in this book, only three routines are needed – addition, subtraction and multiplication of variables and coefficients having pre-defined wordlengths. An Assembly language program is generally less readable which means that the associated documentation must be more carefully written. More care must be exercised in memory management, particularly to avoid problems from multiple use of data memory locations. Assembly directives are available which can (and should!) be used to avoid such problems, but this is not inherent as it is with high-level programming. It is certainly advisable to adopt a more rigorous software design approach if Assembly language is being used, and this may result in a slightly sub-optimum solution. It is however likely to pay dividends in terms of software maintenance and/or upgrading which may become necessary.

Assembly code should be structured in order to enhance its readability. This is most commonly done by using sub-routines, which most engineers understand, but an alternative is to use macros which are in many ways more powerful but are less commonly used, so some comments about these are offered. A sub-routine is a facility provided by the processor, in which a single block of code is located somewhere in memory, terminated by a 'Return' instruction; each time a 'Call' instruction to the sub-routine is encountered, the program flow jumps to this separate block of code, executes it, and returns to the main program after the 'Call' instruction. A macro is a facility provided by the Assembler in the process of generating the machine code [Co81]. A block of code is named and defined by a Macro directive, and whenever the named macro subsequently appears, the defined block of code is inserted by the Assembler. Comparison between sub-routines and macros is usually identified as a trade-off between operating speed and program memory – the 'in-line' code produced when using macros is fast because the program flow is not interrupted by sub-routine calls and returns, but more program memory is needed. However the comparison should go further than this, particularly when considering factors such as structure and readability, because each occurrence of a macro can be individually tailored by means of parameters, which can be included in the macro definition and passed each time the macro is called. This means that they can be made to act just like high-level procedures, a feature which is not possible with sub-routines because the data to be used must be set-up, or pointed to, prior to the call. The only disadvantage of macros is the length of program code, but a judicious combination of macros and sub-routines can be used in the occasional situation where it is a problem. We do not intend to include further details on macros – if you wish to have more background information, see the reference already quoted; the reference list also includes further reading of a more application-specific nature [Cl83, Go89b].

As we have already observed, the results of Chapter 5 show that single-precision operations are unlikely to be satisfactory, even with 16-bit processors. It is therefore necessary to write multi-precision Assembly language routines to add and subtract two variables, and to multiply a variable by a coefficient. The form of these routines will depend upon the precision requirements and upon the basic data wordlength of the processor. It has been pointed out how this might be achieved in general terms, and the large variety of possible processor types and different requirements for precision make it inappropriate to go into too much detail. However, selected Assembly language routines follow for three processors (for which circuits were included in section 6.1). The processor complexity, the precision of the computations and the types of routines are very different and therefore should illustrate a range of possibilities. As in the hardware section, some detail is included (this time in the form of the instruction mnemonics) which can only be fully understood by referring to the appropriate Assembly language reference manual.

(a) *6809 microprocessor*. All three routines, shown as macro definitions in table 6.3, work on 24-bit (3-byte) variables and 8-bit coefficients. The processor offers double-byte instructions for addition and subtraction, and these are made use of in all the routines. It also has an 8 × 8-bit unsigned multiplication which is used three times within the multiplication routine – note that negative variables are correctly handled by subtracting the coefficient from the top byte of the result (as mentioned earlier).

Table 6.3 Arithmetical routines for 6809 microprocessor

```
;*** 3 BYTE ADDITON: VAR1+VAR2 -> VAR1 ***
TRADD    MACRO      VAR1,VAR2
         LDD        VAR1+1    ;Add middle & low bytes
         ADDD       VAR2+1    ;of VAR1 & VAR2,
         STD        VAR1+1    ;and store.
         LDA        VAR1      ;Add high bytes
         ADCA       VAR2      ;of VAR1 & VAR2,
         STA        VAR1      ;and store.
         ENDM

;*** 3 BYTE SUBTRACTION: VAR1-VAR2 -> RSLT ***
SUBTR    MACRO      VAR1,VAR2,RSLT
         LDD        VAR1+1    ;Subtract middle & low bytes
         SUBD       VAR2+1    ;of VAR1 & VAR2,
         STD        RSLT+1    ;and store,
         LDA        VAR1      ;Subtract high bytes
         SBCA       VAR2      ;of VAR1 & VAR2,
         STA        RSLT      ;and store.
         ENDM
;

;*** 3 BYTE MULTIPLICATION: VAR*MANT -> RSLT ***
MULT     MACRO      VAR,MANT,RSLT
         LDA        VAR+2     ;Get low byte,
         LDB        MANT      ;multiply by the
         MUL                  ;mantissa
         STA        RSLT+2    ;and save m.s.byte only.
         LDA        VAR       ;Get the high byte,
         LDB        MANT      ;multiply by
         MUL                  ;the mantissa,
         STD        RSLT      ;and save.
         LDA        VAR+1     ;Get the middle byte
         LDB        MANT      ;and multiply by
         MUL                  ;the mantissa.
         ADDD       RSLT+1    ;Add to middle & low bytes,
         STD        RSLT+1    ;and save.
         LDA        RSLT      ;Add carry
         ADCA       #0        ;to the high byte.
         LDB        VAR       ;Test high byte of variable.
         BPL        POS       ;If negative, subtract the
         SUBA       MANT      ;mantissa from the high byte.
POS:     STA        RSLT      ;Save the high byte.
         ENDM
;
```

(b) *8048 microcontroller*. This device has a very simple instruction set, having no multiplication and not even a subtraction. Six of the device's eight directly-addressable registers are used as two 3-byte working registers for the variables, R2 to R4 being designated as WR1, and R5 to R7 as WR2. The variables are again 24 bit, but this time the coefficient has just four significant bits in its mantissa (which can readily be shown to give 5% accuracy) plus an exponent. Table 6.4 shows the arithmetical routines. Addition is fairly straightforward; subtraction is carried out by negating one variable and adding. Multiplication uses a shift-and-add technique, with the shifted variable only being added into the product if the relevant coefficient bit is 1, which inherently allows for the sign of the variable. The multiplication uses the 3-byte addition macro, and also macros 'RShft1' and 'RShft2' which are not defined, but which perform a single right shift on the variable in WR1 or WR2 in order to divide it by 2. Note that these routines were written for δ-implementation, which for the low accuracy coefficient with a negative exponent is very suitable.

(c) *TMS320E14 Digital Signal Processor*. This is a very powerful device but with a somewhat restricted instruction set. The variables are 32 bit (two word) and the coefficients are 16 bit, and table 6.5 gives the routines. The processor's 16×16 multiplication instruction is very fast, but is signed, which means that the result must be carefully corrected. The detail of this is not appropriate here, but can be appreciated by reference to the device's instruction set. What should be noted is that, although the multiplications have been set-up and executed in the first six instructions, it takes a further fifteen instructions before the result is correct.

High-level vs *low-level software*

Essentially there is a trade-off between the extra hardware cost caused by the inefficiency of high-level software, and the extra design/development cost likely to be associated with programming in Assembly language. The extra hardware cost will be the result of a more sophisticated and expensive processor, greater memory requirements and probably a more expensive printed circuit board. The extra design/development cost will arise because of increased engineering time in order to write, test and document the software.

In a 'one-off' development project, it will probably be cheapest to use a sophisticated processor programmed in a high-level language, and this will also apply in some low-volume applications in which the design costs are a significant proportion of the product's unit cost. In relatively slow-acting control systems, such as chemical processes, a cheap and simple processor, even with the inefficiency of high-level implementation, may be perfectly

Table 6.4 Arithmetic routines for 8048 microcontroller

```
;*** 3 byte addition  WR1 + WR2 --> WR1 ***
TRADD     MACRO
          MOV       A,R7
          ADD       A,R4
          MOV       R4,A        ;Low byte.
          MOV       A,R6
          ADDC      A,R3
          MOV       R3,A        ;Middle byte.
          MOV       A,R5
          ADDC      A,R2
          MOV       R2,A        ;High byte.
          ENDM
;
;*** 3 byte subtraction WR1 - WR2 --> WR1 ***
SUBTR     MACRO
          CLR       C
          CPL       C           ;Set carry (gives +1 for negation of WR2).
          MOV       A,R7
          CPL       A
          ADDC      A,R4
          MOV       R4,A        ;Low byte.
          MOV       A,R6
          CPL       A
          ADDC      A,R3
          MOV       R3,A        ;Middle byte.
          MOV       A,R5
          CPL       A
          ADDC      A,R2
          MOV       R2,A        ;High byte.
          ENDM
;

; *** 3 byte multiplication WR1*MANT*2↑-EXP --> WR1 ***
MULT      MACRO     MANT,EXP
          MOV       R1,#EXP     ;Load exponent.
          MOV       A,#MANT     ;Load mantissa,
          CPL       A           ;complement it
          MOV       R0,A        ;and save.
DIVBY2:   RShft1                ;Divide by
          DJNZ      R1,DIVBY2   ;2↑(EXP+1)
          COPY12                ;WR1 --> WR2.
          RShft+2               ;⎫
          MOV       A,R0        ;⎪ First shift and
          JB6       NOTB6       ;⎬ conditional add.
          TRADD                 ;⎭
NOTB6:    RShft2                ;⎫
          MOV       A,R0        ;⎪ Second shift and
          JB5       NOTB5       ;⎬ conditional add.
          TRADD                 ;⎭
NOTB5:    RShft2                ;⎫
          MOV       A,R0        ;⎪ Third shift and
          JB4       NOTB4       ;⎬ conditional add.
          TRADD                 ;⎭
NOTB4:
          ENDM
```

Table 6.5 Arithmetic routines for TMS 320E14 Digital Signal Processor

```
; *** RES=VAR1+VAR2 ***
Addvar    MACRO     VAR1,VAR2,RES
          ZALS      VAR1+1            ;Load VAR1 into acc.
          ADDH      VAR1
          ADDS      VAR2+1            ;Add VAR2
          ADDH      VAR2
          SACL      RES+1             ;Store in RES.
          SACH      RES
          ENDM
;

; *** RES=VAR1-     VAR2 ***
Subvar    MACRO          VAR1,VAR2,RES
          ZALS      VAR1+1            ;Load VAR1 into acc.
          ADDH      VAR1
          SUBS      VAR2+1            ;Subtract VAR2
          SUBH      VAR2
          SACL      RES+1             ;Store acc. in RES
          SACH      RES
          ENDM

;***      VAR * COEF  --> ACC ***/
Mult      MACRO      VAR,COEF
          LT         VAR+1
          MPY        COEF
          PAC                         ;Low product in P.
          SACH       TEMP+1           ;Save upper word only.
          LT         VAR
          MPY        COEF             ;High product in P.
;
;(Note:   Rest of routine is to correct the result.)
          ZALH       COEF             ;)
          SACL       TEMP             ;)
          BGEZ       CPOS             ;)If COEF is negative,
          ZALH       VAR              ;)then add VAR to
          ADDS       VAR+1            ;)the result.
          ADDS       TEMP+1           ;)
          SACH       TEMP
          SACL       TEMP+1
 CPOS:    ZALH       VAR+1            ;)
          BGEZ       VPOS             ;)If VAR_L is negative, add
          ADDS       COEF             ;)COEF to bottom word of result.
 VPOS:    SUBH       VAR+1
          ADDS       TEMP+1
          ADDH       TEMP
          APAC
          ENDM
```

fast enough (the increased memory requirements are unlikely to be a significant factor).

At the other end of the scale, when applied to very high-speed high-performance controllers, it may be necessary to use low-level programming simply because the performance targets are not achievable otherwise, even with the most sophisticated processor available. In high-volume applications, in which design/development costs often become a very minor proportion of the unit cost, it is probable that hardware costs must be minimised – this requires the cheapest processor and associated memory components, only achievable by programming in Assembly language.

It is also important to remember the other tasks which the software must perform, such as interlocking in the simpler control systems through to adaptation in the more complex. This may drive the decision either way: towards high-level if this is the more appropriate language for the other tasks, or away from high-level if the tasks impose a significant extra burden on the processor. There is no general answer to the 'high-level *vs* low-level?' question, but we have tried to give some idea of the issues involved, and also to identify techniques which reduce the distinction between the two approaches.

Appendix A: Table of Laplace and z-transforms

Laplace transform $f(s)$	Time function $f(t)$ $t > 0$	z-transform $f(z)$	Modified z-transform $f(z,m)$
$\frac{1}{s}$	$u_s(t)$	$\frac{z}{z-1}$	$\frac{1}{z-1}$
$\frac{1}{s^2}$	t	$\frac{Tz}{(z-1)^2}$	$\frac{mT}{(z-1)} + \frac{T}{(z-1)^2}$
$\frac{2}{s^3}$	t^2	$\frac{T^2z(z+1)}{(z-1)^3}$	$T^2\left(\frac{m^2}{(z-1)} - \frac{2m+1}{(z-1)^2} + \frac{2}{(z-1)^3}\right)$
$\frac{1}{s+a}$	e^{-at}	$\frac{z}{z-e^{-aT}}$	$\frac{e^{-amT}}{z-e^{-aT}}$

$\dfrac{1}{(s+a)^2}$	te^{-at}	$\dfrac{Tze^{-aT}}{(z-e^{-aT})^2}$	$\dfrac{Te^{-amT}(e^{-aT}+m(z-e^{-aT}))}{(z-e^{-aT})^2}$
$\dfrac{a}{s(s+a)}$	$1-e^{-at}$	$\dfrac{z(1-e^{-aT})}{(z-1)(z-e^{-aT})}$	$\dfrac{1}{(z-1)}-\dfrac{e^{-amT}}{z-e^{-aT}}$
$\dfrac{1}{(s+a)(s+b)}$	$\dfrac{1}{(b-a)}(e^{-at}-e^{-bt})$	$\dfrac{1}{(b-a)}\left(\dfrac{z}{z-e^{-aT}}-\dfrac{z}{z-e^{-bT}}\right)$	$\dfrac{1}{b-a}\left(\dfrac{e^{-amT}}{z-e^{-aT}}-\dfrac{e^{-bmT}}{z-e^{-bT}}\right)$
$\dfrac{a}{s^2(s+a)}$	$t-\dfrac{1}{a}(1-e^{-at})$	$\dfrac{Tz}{(z-1)^2}-\dfrac{(1-e^{-aT})z}{a(z-1)(z-e^{-aT})}$	$\dfrac{T}{(z-1)^2}+\dfrac{amT-1}{a(z-1)}+\dfrac{e^{-amT}}{a(z-e^{-aT})}$
$\dfrac{a}{s^2+a^2}$	$\sin(at)$	$\dfrac{z\sin(aT)}{z^2-2z\cos(aT)+1}$	$\dfrac{z\sin(amT)+\sin(1-m)aT}{z^2-2z\cos(aT)+1}$
$\dfrac{s}{s^2+a^2}$	$\cos(at)$	$\dfrac{z(z-\cos(aT))}{z^2-2z\cos(aT)+1}$	$\dfrac{z\cos(amT)-\cos(1-m)aT}{z^2-2z\cos(aT)+1}$
$\dfrac{b}{(s+a)^2+b^2}$	$e^{-at}\sin bt$	$\dfrac{z\sin(bT)e^{-aT}}{z^2-2z\cos(bT)e^{-aT}+e^{-2aT}}$	$\dfrac{e^{-amT}(z\sin(bmT)+e^{-aT}\sin(1-m)bT)}{z^2-2z\cos(bT)e^{-aT}+e^{-2aT}}$
$\dfrac{s+a}{(s+a)^2+b^2}$	$e^{-at}\cos bt$	$\dfrac{z(z-\cos(bT)e^{-aT})}{z^2-2z\cos(bT)e^{-aT}+e^{-2aT}}$	$\dfrac{e^{-amT}(z\cos(bmT)+e^{-aT}\cos(1-m)bT)}{z^2-2z\cos(bT)e^{-aT}+e^{-2aT}}$

Appendix B: Block diagram representation of a transfer function

Consider the transfer function

$$F(s) = \frac{n_1 s + n_0}{s^2 + m_1 s + m_0} = \frac{x_0}{x_i} \tag{B.1}$$

and let

$$\frac{y}{x_i} = \frac{1}{s^2 + m_1 s + m_0} \tag{B.2}$$

with

$$\frac{x_0}{y} = n_1 s + n_0 \tag{B.3}$$

Clearly the product of eqns (B.2) and (B.3) equals (B.1).

From (B.2), by transforming into the time domain

$$\ddot{y} = x_i - m_0 y - m_1 \dot{y}$$

so the relationship between x_i and y can be presented as in the diagram of figure B.1.

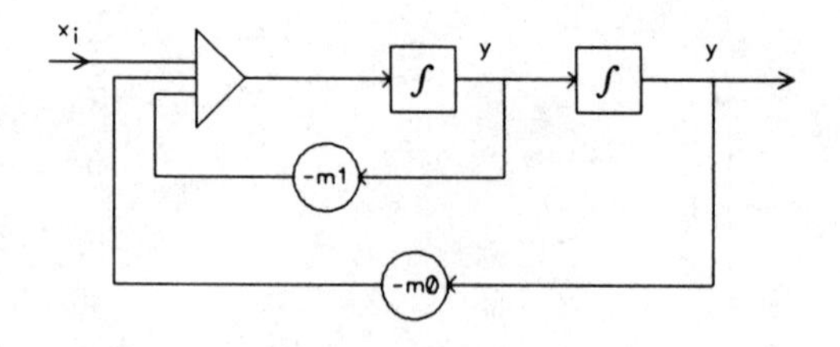

Figure B.1 The denominator of the transfer function

Since, from eqn (B.3)

$$x_o = n_1 \dot{y} + n_o y$$

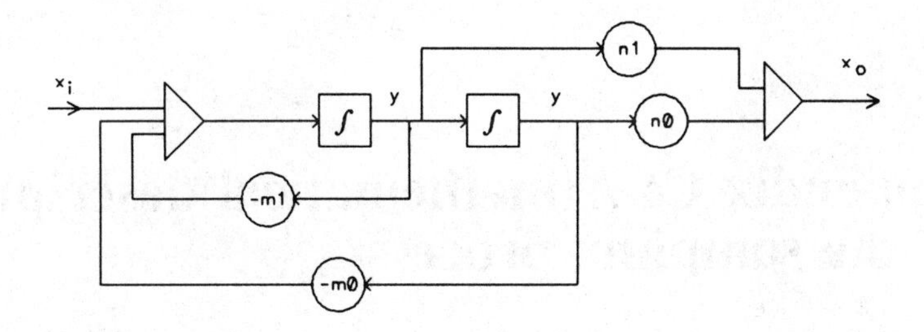

Figure B.2 The complete transfer function

eqn (B.1) may be represented by figure B.2.

Obviously any transfer function can be portrayed in this manner, provided only that the order of the numerator is no greater than that of the denominator (which is the case for any physically realisable system).

Appendix C: A mathematical description of the sampling process

The stream of impulses emerging from the sampler may be related mathematically to the analogue waveform entering it by supposing that the input is multiplied (modulated) by a stream of unit impulses.

To obtain a mathematical description of such a train of impulses, consider first a train of *pulses*. The pulses arrive at intervals of T sec, are of duration α sec, and height H. In a Fourier analysis of this waveform, the fundamental period is T sec so the fundamental frequency is

$$\omega_0 = 2\pi/T \text{ rad/sec}$$

and the waveform is described:

$$x(t) = \tfrac{1}{2}a_0 + a_1 \sin \omega_0 t + a_2 \sin 2\omega_0 t + \ldots$$
$$+ b_1 \cos \omega_0 t + b_2 \cos 2\omega_0 t + \ldots$$

Determining the coefficients a_i and b_i in the usual manner:

(i)

$$\int_0^T x(t)\, dt = \tfrac{1}{2}\, a_0 \int_0^T dt = \tfrac{1}{2}\, a_0 T$$

$$\therefore a_0 = 2H\alpha/T$$

(ii)

$$\int_0^T x(t) \sin \omega_0 t\, dt = a_1 \int_0^T \sin^2 \omega_0 t\, dt = \tfrac{1}{2} a_1 T$$

$$\therefore a_1 = \frac{2}{T}\int_0^T x(t) \sin \omega_0 t\, dt = \frac{2H}{\omega_0 T}(1 - \cos \omega_0 \alpha)$$

Likewise:

$$a_2 = \frac{2H}{2\omega_0 T}(1 - \cos 2\omega_0 \alpha)$$

etc.

(iii)

$$\int_0^T x(t) \cos \omega_0 t \; dt = b_1 \int_0^T \cos^2 \omega_0 t \; dt$$

leading to

$$b_1 = \frac{2H}{\omega_0 T} (\sin \omega_0 \alpha)$$

Likewise:

$$b_2 = \frac{2H}{2\omega_0 T} (\sin 2\omega_0 \alpha)$$

etc.

Thus:

$$x(t) = \frac{H\alpha}{T} + \frac{2H}{\omega_0 T}(1 - \cos \omega_0 \alpha) \sin \omega_0 t + \frac{2H}{2\omega_0 T}(1 - \cos 2\omega_0 \alpha) \sin 2\omega_0 t$$
$$+ \ldots + \frac{2H}{\omega_0 T}(\sin \omega_0 \alpha) \cos \omega_0 t + \frac{2H}{2\omega_0 T}(\sin 2\omega_0 \alpha) \cos 2\omega_0 t + \ldots$$

To create a train of unit impulses, set $H = 1/\alpha$ and let

$$\alpha \to 0$$

$$\sin n\omega_0 \alpha \to n\omega_0 \alpha$$

$$\cos n\omega_0 \alpha \to 1$$

giving

$$x(t) = \frac{1}{T} + \frac{2}{T} \cos \omega_0 t + \frac{2}{T} \cos 2\omega_0 t + \ldots$$

$$= \frac{1}{T} + \frac{2}{T} \sum_{n=1}^{\infty} \cos n\omega_0 t$$

$$= \frac{1}{T} + \frac{2}{T} \sum_{n=1}^{\infty} \tfrac{1}{2} [e^{jn\omega_0 t} + e^{-jn\omega_0 t}]$$

$$= \frac{1}{T} \sum_{n=-\infty}^{\infty} e^{jn\omega_0 t} = \frac{1}{T} \sum_{n=-\infty}^{\infty} e^{-jn\omega_0 t} \tag{C.1}$$

If in general terms the input to an impulse sampler is denoted $f(t)$, then it is the convention to denote the output by $f^*(t)$. Thus

$$f^*(t) = f(t)x(t)$$

and

$$\mathcal{L}\{f^*(t)\} = f^*(s) = \int_0^\infty f^*(t)\, e^{-st}\, dt$$

After substituting for $x(t)$ from eqn (C.1):

$$f^*(s) = \int_0^\infty f(t) \sum_{-\infty}^{\infty} \frac{1}{T} e^{-(s+jn\omega_0)t}\, dt$$

$$= \frac{1}{T} \sum_{-\infty}^{\infty} f(s + jn\omega_0) \tag{C.2}$$

This relationship between $f(s)$ and $f^*(s)$ has sometimes been used in the design of sampled data systems [Sl74] to given an approximation to the frequency response of the sampled system by summing a number of terms of the infinite series, perhaps as many as forty.

More importantly, it can be seen that because of the summation in eqn (C.2) over an infinite range of n, the frequency content of the original signal described by $f(s)$ is repeated at intervals over the complete spectrum to form $f^*(s)$. This ties up with the material of Chapter 2 where the primary strip is described.

The pulse sampler

An alternative view of the sampling process is possible. When the waveform $f(t)$ is sampled, the result is not, according to this model, a sequence of impulses but of *pulses* having the height of $f(t)$ at the sampling instant and a duration α that is small but finite. The sampled waveform $f^*_\alpha(t)$ is now regarded as the product of $f(t)$ and a train $p(t)$ of pulses of unit height and duration α. Because α is considered very small bearing in mind the rates of change in $f(t)$, the series of pulses that constitutes $f^*_\alpha(t)$ may be taken as rectangular.

In mathematical terms:

$$f^*_\alpha(t) = \sum_{n=0}^{\infty} f(kT)[u_s(t - kT) - u_s(t - kT - \alpha)]$$

Taking Laplace transforms:

$$f^*_\alpha(s) = \sum_{n=0}^{\infty} f(kT) \left[\frac{1}{s} e^{-kTs} - \frac{1}{s} e^{-(kT+\alpha)s}\right]$$

$$= \sum_{n=0}^{\infty} f(kT)\,\frac{1}{s}\,[1 - e^{-\alpha s}]e^{-kTs}$$

Since $e^{-\alpha s} = 1 - \alpha s + \alpha^2 s^2/2 - \ldots \simeq 1 - \alpha s$:

$$f_{\alpha}^{*}(s) = \sum_{n=0}^{\infty} f(kT)\,\frac{1}{s}\,[\alpha s]e^{-kTs}$$

$$= \alpha \sum_{n=0}^{\infty} f(kT)e^{-kTs}$$

Because of the reformulation of the sampling process, we must also reform the description of the ZOH. This hitherto involved the integration of the impulse of strength $f(kT)$ to produce a step of height $f(kT)$; now we have a pulse of strength $\alpha f(kT)$, the integral of which is approximately a step but of height $\alpha f(kT)$. To produce the necessary result we must then divide by α, so the output of the ZOH is described in this case by

$$x_g(kT) = [\alpha f(kT)]\left[\frac{1}{s}(1 - e^{-sT})\,\frac{1}{\alpha}\right]$$

$$= f(kT)\left[\frac{1}{s}(1 - e^{-sT})\right] \qquad \text{(C.3)}$$

Eqn (C.3) clearly gives the same expression for x_g as that developed earlier, so either model of the sample-and-hold operation is acceptable, but the impulse sampler is the more frequently encountered.

Appendix D: Compensation techniques

In this appendix, two much used compensation techniques are described in s-domain terms; they are employed in several examples of w-plane design and are given below for the benefit of those readers who are unfamilar with the detail of the design processes. Both techniques make use of the Nichols Chart.

There are three ways of displaying a frequency response: as a Bode plot, as a Nyquist diagram, and as a Nichols diagram. Each of these has its *raison d'être*, and in the case of the Nichols diagram it is the ease with which we can see the effect of open loop gain changes on the closed loop frequency response; this is particularly relevant when designing compensation for a control system.

There is a unique relationship between the open loop transfer function $H(s)$ and the closed loop transfer function $F(s)$, namely

$$F(s) = \frac{H(s)}{1 + H(s)}$$

assuming negative feedback. Thus $F(s)$ or $F(j\omega)$ is completely determined when we know $H(s)$ or $H(j\omega)$. To use a Nichols Chart, $H(j\omega)$ is plotted on the gain/phase plane, that is, the gain of the magnitude of $H(j\omega)$, for a number of values of ω, is plotted against the phase. Superimposed on this locus are the contours relating $F(j\omega)$ to $H(j\omega)$ for all possible values of $H(j\omega)$, so one can read off by inspection the gain and phase of $F(j\omega)$ at any given value of ω, from a plot of $H(j\omega)$.

D.1 The Phase advance compensator

Plants that are insufficiently stable are often stabilised to the required degree by the use of a phase advance (or phase lead) network. The design of such a compensator may be undertaken in a number of ways but perhaps the easiest is that carried out on the Nichols Chart. The design process makes use of three simple facts about the network which are given below as eqns (D.2), (D.3) and (D.4).

The transfer function of a phase lead network takes the form:

$$C(s) = G\left[\frac{1 + skT_c}{1 + sT_c}\right] \qquad k > 1 \tag{D.1}$$

Its most important characteristic is the shape of the phase/frequency curve, shown in figure D.1, where it can be seen that over a range of frequencies the phase is 'advanced', with the maximum advance ϕ_c occurring at the so-called 'centre frequency', ω_c. The magnitude of ϕ_c is

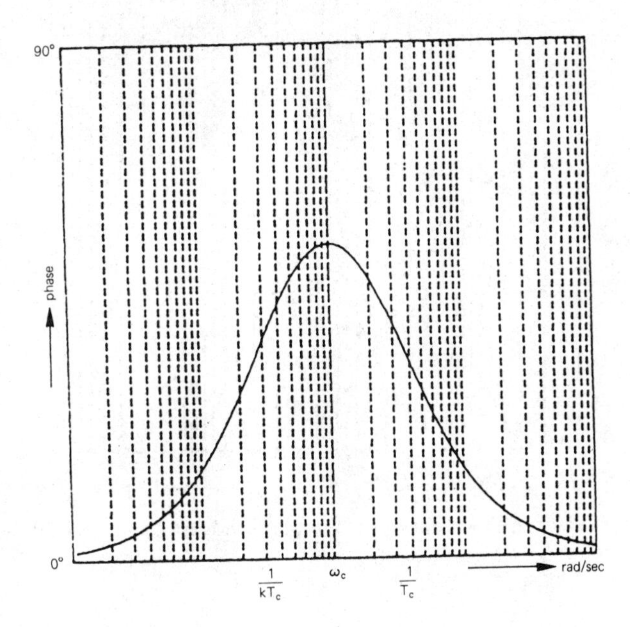

Figure D.1 Phase/frequency curve for the phase advance network, shown for $T_c = 0.316$ sec, $k = 10$

determined by the choice of k, while the frequency at which it occurs, ω_c, is fixed by both k and T_c. The relationships between these parameters are simple in form and easily remembered; they are derived below, but stated here for convenience:

$$\sin \varphi_c = \frac{k - 1}{k + 1} \tag{D.2}$$

$$\omega_c = \frac{1}{T_c\sqrt{k}} \tag{D.3}$$

The third relationship concerns the gain/frequency characteristic of $C(j\omega)$ which is shown in figure D.2, for $G = 1$. At the centre frequency ω_c, the gain is $\sqrt{k}$. In general, therefore (that is, for $G \neq 1$), the gain of $C(j\omega_c) = M_c$ where

$$M_c = G\sqrt{k} \tag{D.4}$$

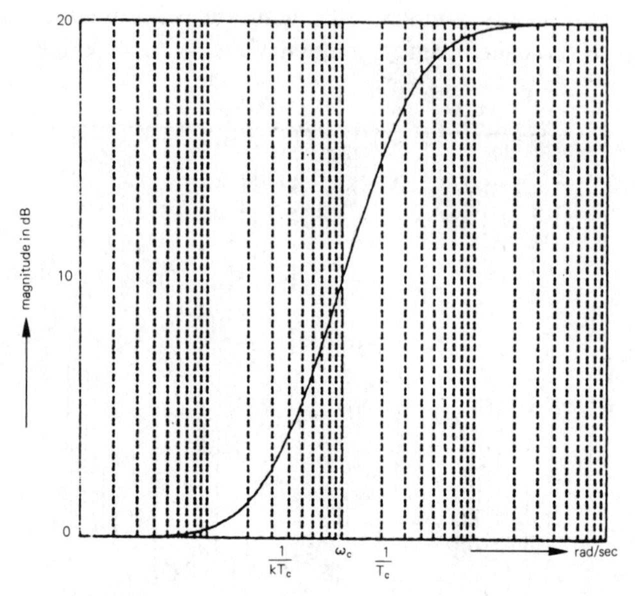

Figure D.2 Gain/frequency curve for the phase advance network, shown for $T_c = 0.316$ sec, $k = 10$

The three equations (D.2) to (D.4) are used in a simple design procedure that is summarised in the following steps:

(1) Choose a point P on the Nichols Chart through which you would like the compensated frequency response to pass.
This might, for instance, lie on the +3 dB contour as in figure D.3 (point P_1), where the intention is that the compensated curve should touch (that is, make a tangent with) the contour at this point; the *closed loop* frequency response would therefore have a *peak* gain of 3 dB. There is no way of ensuring that your choice of position for P will indeed lead to a *tangent* with the contour – you are making an educated guess – but if the final result is unsatisfactory, the process may easily be repeated.
A common alternative to designing for a given peak closed loop gain is to design for a suitable phase margin, usually 40° to 50°. In that case,

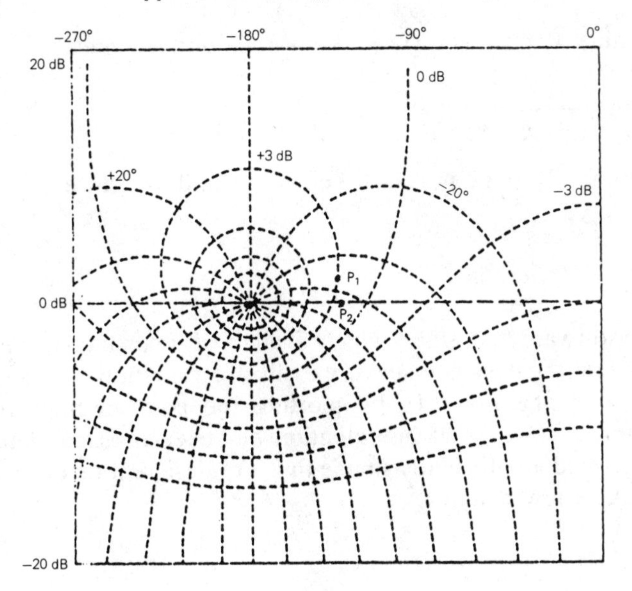

Figure D.3 The Nichols Chart. The closed loop gain contours are shown at intervals of 3 dB, the phase contours at intervals of 20°

the point P would lie on the zero dB line, 40° or 50° from the critical point (figure D.3, point P_2).

(2) Decide upon the frequency ω_p that the compensated contour should have at the point P.
This is obviously the frequency of the peak closed loop gain in the compensated system in the case where $P = P_1$, and in the other case ($P = P_2$) it is the crossover frequency. The crossover frequency gives a good indication of the speed of response of your system; the rise time in the step response would be something around half a period at the crossover frequency.

(3) Locate the point R on the *uncompensated* frequency response with the same frequency as P (that is, ω_p).
The compensating network will so alter the frequency response curve that the point R moves to P. Note the horizontal and vertical translations, φ and M, needed to do this.

(4) Equate the value chosen for ω_p to the centre frequency of the compensator, that is, $\omega_p = \omega_c$.
The horizontal translation φ in (3) is therefore equal to φ_c in eqn (D.2) and the vertical translation M is M_c in eqn (D.4).

(5) Determine k from eqn (D.2), knowing φ_c, since

$$k = \frac{1 + \sin \varphi_c}{1 - \sin \varphi_c}$$

(6) Determine T_c from eqn (D.3), knowing k and ω_c, since

$$T_c = \frac{1}{\omega_c \sqrt{k}}$$

(7) Determine G from eqn (D.4), knowing k and M_c, since

$$G = M_c / \sqrt{k}$$

Note that M_c here is *not* in dB.

With the software that is now widely available for personal computers, it is probable that the designer can very easily ascertain how well he guessed the position of the point P, by plotting the compensated frequency response and checking that his intention has been realised. The design process described is sufficiently simple that a re-design, if that is necessary, will only take a few minutes.

Derivations

The phase/frequency diagram is shown in figure D.1 and it is clear from the symmetry that the maximum value φ_c will occur at ω_c, halfway between $1/T_c$ and $1/T_c k$.

If

$$x_1 = \log(1/T_c k)$$

and

$$x_2 = \log(1/T_c)$$

then x_c, midway between x_1 and x_2, is given by:

$$\begin{aligned} x_c &= \tfrac{1}{2}(x_1 + x_2) \\ &= \tfrac{1}{2}\left(\log \frac{1}{T_c k} + \log \frac{1}{T_c}\right) \\ &= \tfrac{1}{2}\left(\log \frac{1}{T_c^2 k}\right) \\ &= \log \frac{1}{T_c \sqrt{k}} \\ &= \log \omega_c \end{aligned}$$

that is

$$w_c = \frac{1}{T_c\surd k} \qquad \text{(D.3)}$$

At this frequency, 'the centre frequency':

$$C(j\omega_c) = G\left[\frac{1 + j\omega_c T_c k}{1 + j\omega_c T_c}\right]$$

Substituting for ω_c from eqn (D.3):

$$C(j\omega_c) = G\left[\frac{1 + j\surd k}{1 + j(1\surd k)}\right]$$

so the phase of the network at ω_c is

$$\varphi_c = \alpha - \beta$$

where $\alpha = \tan^{-1}(\surd k)$, $\beta = \tan^{-1}(1/\surd k)$.

Hence $\sin \varphi_c = \sin \alpha \cos \beta - \sin \beta \cos \alpha$

$$= \frac{k - 1}{k + 1} \quad \text{(see figure D.4)} \qquad \text{(D.2)}$$

at $\omega = \omega_c$.

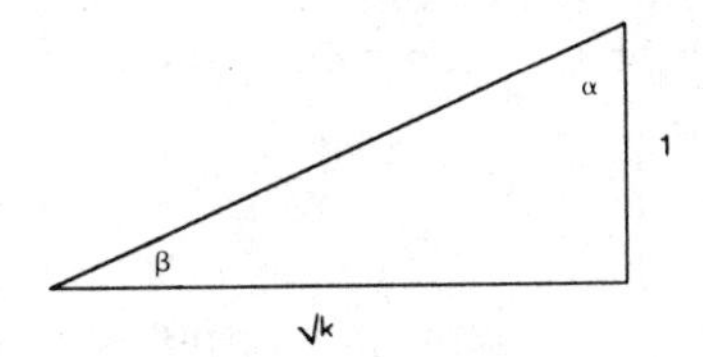

Figure D.4 Geometrical relationships

The magnitude at this frequency is

$$M_c = G\frac{(1 + k)^{1/2}}{\left(1 + \dfrac{1}{k}\right)^{1/2}} = G\surd k$$

It would appear in theory as if any phase advance up to 90° is possible with this network, but in fact the maximum is more like 60° in practice. The limit is imposed by the existence of noise in the system, which is of course inevitable. Figure D.2 shows that high-frequency signals (that is, noise) will be amplified by a factor of k while the signal frequencies of interest will only be amplified by about $\sqrt{k}$: thus the signal-to-noise ratio will be reduced by a factor of around $\sqrt{k}$. The reduction that can be tolerated must depend upon circumstances, upon the amount of noise present, but it is likely that a maximum for k will lie in the range 10–20; the maximum value of φ_c will therefore lie between 55° and 65°

D.2 The Proportional plus Integral compensator

Many plants, particularly in process control, suffer from an excess of stability – they are too sluggish in their time response. It is characteristic of such processes that they can often be approximated to first- or second-order (or sometimes higher) plants with purely real poles, that is, to

$$G_1(s) = \frac{K_p}{(1 + sT_1)} \tag{D.5}$$

or

$$G_2(s) = \frac{K_p}{(1 + sT_1)(1 + sT_2)} \tag{D.6}$$

The sluggish response in such cases is frequently dealt with using the Proportional plus Integral or PI Controller.

The transfer function of a PI controller takes the form

$$C(s) = K\left[\frac{1 + sT_c}{sT_c}\right] \tag{D.7}$$

It may be used to alter the frequency response of a first-order plant in the way shown in figure D.5, or a second-order plant as shown in figure D.6.

The essential mathematical relationships are derived as follows. From eqn (D.7):

$$C(j\omega) = \frac{K}{T_c} \cdot \frac{1 + j\omega T_c}{j\omega}$$

Hence

$$\left| C(j\omega) \right| = M = \frac{K(1 + \omega^2 T_c^2)^{1/2}}{\omega T_c} \tag{D.8}$$

and

$$\angle C(j\omega) = \tan^{-1}\omega T_c - 90° = -\varphi$$

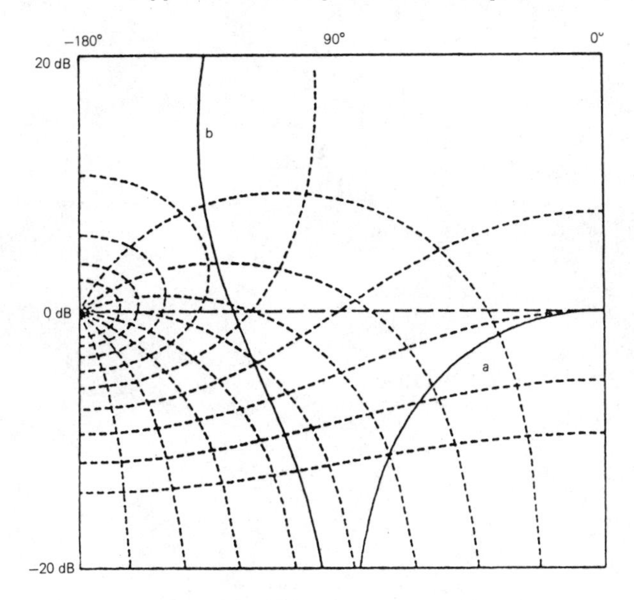

Figure D.5 Action of a PI controller on the frequency response of a first-order plant: (a) plant alone; (b) plant and controller

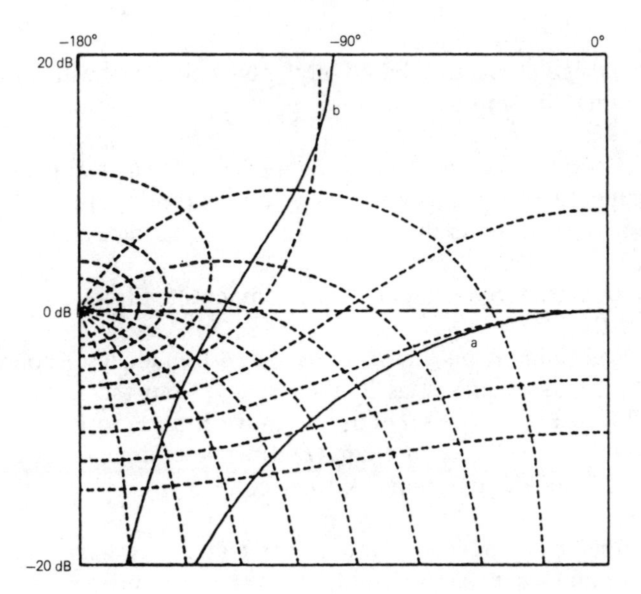

Figure D.6 Action of a PI controller on the frequency response of a second-order plant: (a) plant alone; (b) plant and controller

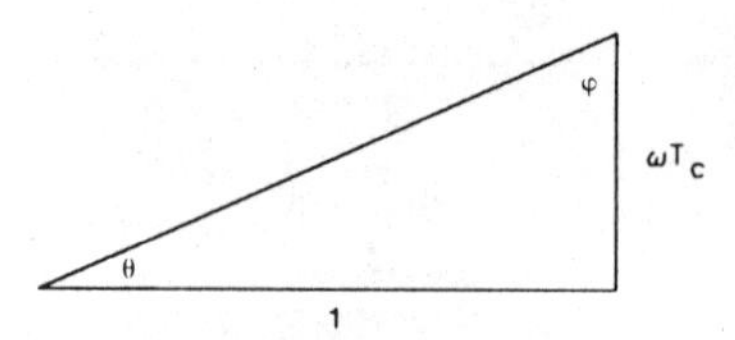

Figure D.7 Geometrical relationships

that is

$$\omega T_c = \tan(90° - \varphi) = \tan\theta$$

where $\theta = 90° - \varphi$.

Thus

$$T_c = \frac{\tan\theta}{\omega} \tag{D.9}$$

The trigonometry is illustrated in figure D.7, from which it can be seen that

$$\sin\theta = \omega T_c/(1 + \omega^2 T_c^2)^{1/2}$$

Hence, from eqn (D.8):

$$K = M\sin\theta \tag{D.10}$$

A simple design procedure, based on the Nichols Chart with eqns (D.9) and (D.10), may be set out as follows.

(1) Choose a point P on the Nichols Chart through which you would like the compensated frequency response to pass (see figure D.8).
The same factors affect the choice of P here as in the case of the phase advance compensator.
(2) Decide upon the frequency ω_p of the compensated frequency response curve at P.
(3) Locate the point R on the *uncompensated* frequency response curve with the same frequency as P, that is, ω_p. Note the horizontal and vertical travel (φ and M) required to move from R to P.
(4) Given that $\theta = 90° - \varphi$, use eqn (D.9) to determine T_c for $\omega = \omega_p$.
(5) Knowing θ and M, determine K from eqn (D.10).

As with the phase advance network, it may require more than one attempt to obtain a satisfactory result, but the design process is so simple that this should present no difficulty.

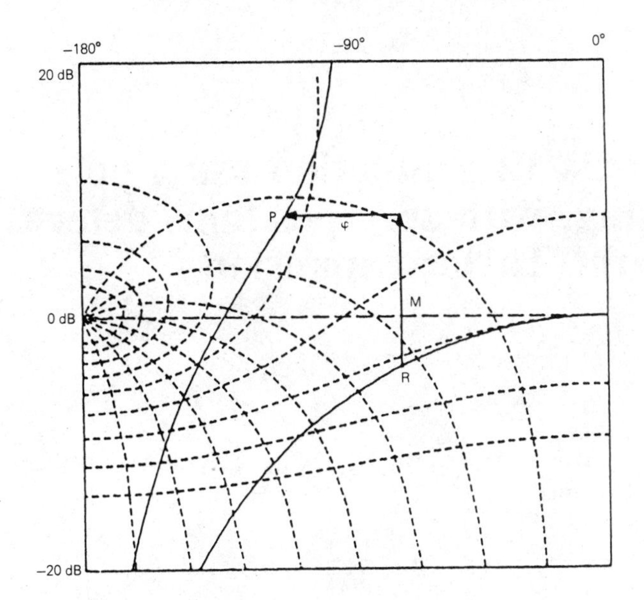

Figure D.8 Determination of φ and M in the design of a PI controller

Appendix E: Emulation using pole mapping, with zero positions determined using the Taylor expansion

Suppose we wish to transform $G(s)$ into the z-domain, creating $G(z)$, where for example

$$G(s) = \frac{n_0 + n_1 s}{m_0 + m_1 s + m_2 s^2} = \frac{y(s)}{u(s)} \tag{E.1}$$

E.1 The basic method

We begin by determining the pole positions of $G(s)$ and translating these directly into the z-plane, by use of the mapping $z = e^{sT}$, to form the poles of $G(z)$. Let

$$G(z) = \frac{a_0 + a_1 z^{-1} + a_2 z^{-2}}{1 + b_1 z^{-1} + b_2 z^{-2}} = \frac{y(z)}{u(z)} \tag{E.2}$$

The coefficients b_i are known from the mapping, that is

$$1 + b_1 z^{-1} + b_2 z^{-2} = (1 - \beta_1 z^{-1})(1 - \beta_2 z^{-1})$$

where the β_i are the mapped s-plane poles.

The first step in determining the zeros of $G(z)$ is to note that the eqns (E.1) and (E.2) may be re-written in the time domain as

$$n_o u(t) + n_1 \overset{(1)}{u}(t) = m_0 y(t) + m_1 \overset{(1)}{y}(t) + m_2 \overset{(2)}{y}(t) \tag{E.3}$$

and

$$a_0 u(t) + a_1 u(t - T) + a_2 u(t - 2T) = y(t) + b_1 y(t - T) + b_2 y(t - 2T) \tag{E.4}$$

Since terms such as $u(t - T)$ etc. may be expanded in Taylor series, eqn (E.4) can be re-stated in derivatives rather than sample values:

$$u(t)[a_0 + a_1 + a_2] - \overset{(1)}{u}(t)T\,[a_1 + 2a_2] + \overset{(2)}{u}(t)\,\frac{T}{2!}\,[a_1 + 4a_2] - \ldots$$

$$= y(t)[1 + b_1 + b_2] - \overset{(1)}{y}(t)T\,[b_1 + 2b_2] + \overset{(2)}{y}(t)\,\frac{T}{2!}\,[b_1 + 4b_2] - \ldots \qquad \text{(E.5)}$$

We now have two equations in derivatives of $u(t)$ and $y(t)$, one of which (E.5) is supposed to represent the other (E.4), to some degree of approximation, through a proper choice of the a-coefficients. (Remember that the b-coefficients are already specified.) The obvious step of equating terms in the two equations will not work – there are fewer equations than unknown quantities for example. The idea can be made to work, however, if eqn (E.1) is modified to

$$\hat{G}(s) = \frac{m_0 + n_1 s}{m_0 + m_1 s + m_2 s^2} \cdot \frac{f_0 + f_1 s + f_2 s^2}{f_0 + f_1 s + f_2 s^2} = \frac{y(s)}{u(s)} \qquad \text{(E.6)}$$

where the coefficients f_i of the auxiliary function $F(s) = f_0 + f_1 s + f_2 s^2$ are treated as unknowns that need to be determined along with the a_i. In the general case, the number of f-coefficients is chosen to equal the number of a-coefficients.

Eqn (E.6) is now expressed in the time domain like eqn (E.3) and equated term by term with eqn (E.5), to give a result that, expressed in matrix form, may be written:

$$\begin{bmatrix} 1 & 1 & 1 \\ 0 & -T & -2T \\ 0 & T^2/2! & 4T^2/2! \end{bmatrix} \begin{bmatrix} 1 \\ b_1 \\ b_2 \end{bmatrix} = \begin{bmatrix} m_0 & 0 & 0 \\ m_1 & m_0 & 0 \\ m_2 & m_1 & m_0 \end{bmatrix} \begin{bmatrix} f_0 \\ f_1 \\ f_2 \end{bmatrix} \qquad \text{(E.7)}$$

$$\begin{bmatrix} 1 & 1 & 1 \\ 0 & -T & -2T \\ 0 & T^2/2! & 4T^2/2! \end{bmatrix} \begin{bmatrix} a_0 \\ a_1 \\ a_2 \end{bmatrix} = \begin{bmatrix} n_0 & 0 & 0 \\ n_1 & n_0 & 0 \\ 0 & n_1 & n_0 \end{bmatrix} \begin{bmatrix} f_0 \\ f_1 \\ f_2 \end{bmatrix} \qquad \text{(E.8)}$$

Before considering the solution of these equations, note the obvious forms or patterns in the matrices. These are important because they allow us to write down by inspection the necessary equations for any case. If, for example, we wished to formulate $G(z)$ with an extra term in the numerator

$$G(z) = \frac{a_0 + a_1 z^{-1} + a_2 z^{-2} + a_3 z^{-3}}{1 + b_1 z^{-1} + b_2 z^{-2}}$$

then the appropriate equations would have four f-coefficients, since there are now four a-coefficients, and would be

$$\begin{bmatrix} 1 & 1 & 1 \\ 0 & -T & -2T \\ 0 & T^2/2! & 4T^2/2! \\ 0 & -T^3/3! & -8T^3/3! \end{bmatrix} \begin{bmatrix} 1 \\ b_1 \\ b_2 \end{bmatrix} = \begin{bmatrix} m_0 & 0 & 0 & 0 \\ m_1 & m_0 & 0 & 0 \\ m_2 & m_1 & m_0 & 0 \\ 0 & m_2 & m_1 & m_0 \end{bmatrix} \begin{bmatrix} f_0 \\ f_1 \\ f_2 \\ f_3 \end{bmatrix}$$

and

$$\begin{bmatrix} 1 & 1 & 1 & 1 \\ 0 & -T & -2T & -3T \\ 0 & T^2/2! & 4T^2/2! & 9T^2/2! \\ 0 & -T^3/3! & -8T^3/3! & -27T^3/3! \end{bmatrix} \begin{bmatrix} a_0 \\ a_1 \\ a_2 \\ a_3 \end{bmatrix} =$$

$$\begin{bmatrix} n_0 & 0 & 0 & 0 \\ n_1 & n_0 & 0 & 0 \\ 0 & n_1 & n_0 & 0 \\ 0 & 0 & n_1 & n_0 \end{bmatrix} \begin{bmatrix} f_0 \\ f_1 \\ f_2 \\ f_3 \end{bmatrix}$$

If $G(s)$ and $G(z)$ were simpler, say

$$G(s) = \frac{n_0 + n_1 s}{m_0 + m_1 s}$$

$$G(z) = \frac{a_0 + a_1 z^{-1}}{1 + b_1 z^{-1}}$$

the equations would be:

$$\begin{bmatrix} 1 & 1 \\ 1 & -T \end{bmatrix} \begin{bmatrix} 1 \\ b_1 \end{bmatrix} = \begin{bmatrix} m_0 & 0 \\ m_1 & m_0 \end{bmatrix} \begin{bmatrix} f_0 \\ f_1 \end{bmatrix}$$

and

$$\begin{bmatrix} 1 & 1 \\ 0 & -T \end{bmatrix} \begin{bmatrix} a_0 \\ a_1 \end{bmatrix} = \begin{bmatrix} n_0 & 0 \\ n_1 & n_0 \end{bmatrix} \begin{bmatrix} f_0 \\ f_1 \end{bmatrix}$$

It happens quite often that either the numerator or the denominator of $G(s)$ has a leading zero, for example, $m_0 = 0$. In that case, the f-coefficients cannot be determined by eqn (E.7) as it is shown, and it is necessary to alter it to:

$$\begin{bmatrix} 0 & -T & -2T \\ 0 & T^2/2! & 4T^2/2! \\ 0 & -T^3/3! & -8T^3/3! \end{bmatrix} \begin{bmatrix} 1 \\ b_1 \\ b_2 \end{bmatrix} = \begin{bmatrix} m_1 & 0 & 0 \\ m_2 & m_1 & 0 \\ 0 & m_2 & m_1 \end{bmatrix} \begin{bmatrix} f_0 \\ f_1 \\ f_2 \end{bmatrix} \tag{E.9}$$

This is done because there is no equation relating coefficients of $y(t)$ when $m_0 = 0$, so the three f-coefficients must be determined by equating coefficients of $\overset{(1)}{y}(t)$, $\overset{(2)}{y}(t)$, and $\overset{(3)}{y}(t)$.

It is worth considering the solution of eqns (E.7) and (E.8) when (a) it is undertaken by a computer program and (b) undertaken by hand.

(a) The programmed solution

Eqns (E.7) and (E.8) may be written more briefly as

$$CB = MF \tag{E.10}$$

and

$$VA = NF \tag{E.11}$$

where A, B and F are the column vectors of a-, b- and f-coefficients, and C, M, V and N are defined by reference to eqns (E.7) (or (E.9)) and (E.8). At the moment, the matrices C and V are seen to be identical, but they are not necessarily so, as will be shown below, and it is therefore necessary to differentiate between them.

Eliminating F from eqns (E.10) and (E.11) gives

$$A = V^{-1}NM^{-1}CB \tag{E.12}$$

The matrices V, N, M and C may all be written down by inspection, and the inverses of V and M will always exist. B is readily computed from the poles of $G(s)$ so the evaluation of the A-coefficients from eqn (E.12) is quite simple.

Determining A through the use of the bilinear transform is conceptually a less complex process, but it is very tedious to carry out in all but the simplest cases. It, too, is therefore best done by computer program, but the reader may well find that the computation in eqn (E.12) is easier to code up than the substitution of the bilinear transform.

It is worth remarking that the same pair of equations, (E.7) and (E.8) or (E.10) and (E.11), can be used to determine $G(s)$ given $G(z)$ – the inverse transformation. In that case M is determined from B by pole mapping and N is determined from eqn (E.11). The latter step requires the observation that

$$\begin{bmatrix} n_0 & & \\ n_1 & n_0 & \\ n_2 & n_1 & n_0 \end{bmatrix} \begin{bmatrix} f_0 \\ f_1 \\ f_2 \end{bmatrix} = \begin{bmatrix} f_0 & & \\ f_1 & f_0 & \\ f_2 & f_1 & f_0 \end{bmatrix} \begin{bmatrix} n_0 \\ n_1 \\ n_2 \end{bmatrix}$$

(b) Manual solution

This is best carried out by explicitly determining the *f*-coefficients from eqn (E.7) since the triangular form of M makes this easy.

In eqn (E.8), note that

$$V = \begin{bmatrix} 1 & & \\ & T & \\ & & T^2/2! \end{bmatrix} \begin{bmatrix} 1 & 1 & 1 \\ 0 & -1 & -2 \\ 0 & 1 & 4 \end{bmatrix}$$

so

$$A = \begin{bmatrix} 1 & 1 & 1 \\ 0 & -1 & -2 \\ 0 & 1 & 4 \end{bmatrix}^{-1} \begin{bmatrix} 1 & & \\ & 1/T & \\ & & 2!/T^2 \end{bmatrix} [N][F] \qquad \text{(E.13)}$$

The only problem in determining A is the matrix inversion in eqn (E.13) but there is a simple way around it. The matrix to be inverted, call it H, is of dimensions $n \times n$, where n is the number of a-coefficients to be evaluated; the matrix always has the same form, whatever the problem; only the dimension varies. Since the order of a controller is unlikely to be larger than three or four (coefficient sensitivity would alone ensure this), one can tabulate H^{-1} for $n = 2, 3, 4$. These are designated H_n^{-1} and are given in table E.1.

Table E.1 Values of H_n^{-1} for n up to 4

$$H_2^{-1} = \begin{bmatrix} 1 & 1 \\ 0 & -1 \end{bmatrix} \qquad H_3^{-1} = \frac{1}{2!}\begin{bmatrix} 2 & 3 & 1 \\ 0 & -4 & -2 \\ 0 & 1 & 1 \end{bmatrix}$$

$$H_4^{-1} = \frac{1}{3!}\begin{bmatrix} 6 & 11 & 6 & 1 \\ 0 & -18 & -15 & -3 \\ 0 & 9 & 12 & 3 \\ 0 & -2 & -3 & -1 \end{bmatrix}$$

To illustrate how easily the method can be applied, whatever the intricacies of its derivation, consider the problem of determining the coefficients for a three-point integration algorithm, that is

$$G(s) = \frac{1}{s}$$

and

$$G(z) = \frac{a_0 + a_1 z^{-1} + a_2 z^{-2}}{1 + b_1 z^{-1}}$$

This is an example in which the denominator of $G(s)$ has a leading zero ($m_0 = 0$) so the relevant equations are (E.9) and (E.8) (or (E.13)), suitably amended.

Thus, the pole of $G(s)$ is at $s = 0$, leading to the pole of $G(z)$ at $z = 1$, so $b_1 = -1$. Hence

$$\begin{bmatrix} 0 & -T \\ 0 & T^2/2! \\ 0 & -T^3/3! \end{bmatrix} \begin{bmatrix} -1 \\ \\ 1 \end{bmatrix} = \begin{bmatrix} 1 & & \\ & 1 & \\ & & 1 \end{bmatrix} \begin{bmatrix} f_0 \\ f_1 \\ f_2 \end{bmatrix}$$

giving $f_0 = T$
$f_1 = -T^2/2!$
$f_2 = T^3/3!$

Substituting in eqn (E.13)

$$\begin{bmatrix} a_0 \\ a_1 \\ a_2 \end{bmatrix} = \begin{bmatrix} 1 & 1 & 1 \\ 0 & -1 & -2 \\ 0 & 1 & 4 \end{bmatrix}^{-1} \begin{bmatrix} 1 & & \\ & 1/T & \\ & & 2!/T^2 \end{bmatrix} \begin{bmatrix} T \\ -T^2/2! \\ T^3/3! \end{bmatrix}$$

$$= \frac{1}{2!} \begin{bmatrix} 2 & 3 & 1 \\ 0 & -4 & -2 \\ 0 & 1 & 1 \end{bmatrix} \begin{bmatrix} T \\ -T/2 \\ T/3 \end{bmatrix}$$

$$= \frac{T}{12} \begin{bmatrix} 5 \\ 8 \\ -1 \end{bmatrix}$$

The result is therefore

$$G(z) = \frac{T}{12} \cdot \frac{5 + 8z^{-1} - z^{-2}}{1 - z^{-1}}$$

This is in fact quite an elegant way to derive the well known Adams–Bashforth algorithm, AB3, which this is.

E.2 An enhancement

Up to this point, the quantity to be calculated has been the value of the output variable at the time-instant of the latest input sample. However, there is no reason why the value of the output should not be calculated for some other instant of time from the same input samples; that instant, referred to as the 'pivot', may be in the future or in the past (with respect to the latest input sample).

This may be accomplished very readily by rewriting eqn (E.4) as:

$$\begin{aligned} & a_0u(t) + a_1u(t - T) + a_2u(t - 2T) \\ & = y(t + DT) + b_1y(t + DT - T) + b_2y(t + DT - 2T) \end{aligned} \tag{E.14}$$

where D is the time shift of the pivotal point and is stated in multiples of the sampling interval. It does not have to be an integer and may be positive or negative ('positive' meaning, into the future).

The effect on the matrix formulation of the equations is to alter C (eqn (E.7)) to:

$$C = \begin{bmatrix} 1 & 1 & 1 \\ DT & (D-1)T & (D-2)T \\ (DT)^2/2! & (D-1)^2T^2/2! & (D-2)^2T^2/2! \end{bmatrix} \tag{E.15}$$

Otherwise there is no change.

To illustrate the use of the enhanced formulation, let us derive an algorithm to predict the value of the next sample using the latest four samples. The 'transfer function' in question is simply

$$G(s) = 1/1$$

that is, $n_0 = 1$, $m_0 = 1$. With no poles in $G(s)$, there are none in $G(z)$ so $b_i = 0$; because we are predicting the next value, $D = +1$. Thus

$$\begin{bmatrix} 1 \\ DT \\ \dfrac{(DT)^2}{2!} \\ \dfrac{(DT)^3}{3!} \end{bmatrix} = \begin{bmatrix} f_0 \\ f_1 \\ f_2 \\ f_3 \end{bmatrix} = \begin{bmatrix} 1 \\ T \\ T^2/2! \\ T^3/3! \end{bmatrix}$$

From eqn (E.8), suitably extended:

$$\begin{bmatrix} a_0 \\ a_1 \\ a_2 \\ a_3 \end{bmatrix} = \frac{1}{3!} \begin{bmatrix} 6 & 11 & 6 & 1 \\ 0 & -18 & -15 & -3 \\ 0 & 9 & 12 & 3 \\ 0 & -2 & -3 & -1 \end{bmatrix} \begin{bmatrix} 1 & & & \\ & 1/T & & \\ & & 2!/T^2 & \\ & & & 3!/T^2 \end{bmatrix} \begin{bmatrix} 1 \\ T \\ T^2/2! \\ T^2/3! \end{bmatrix}$$

$$\therefore \begin{bmatrix} a_0 \\ a_1 \\ a_2 \\ a_3 \end{bmatrix} = \frac{1}{3!} \begin{bmatrix} 6 & 11 & 6 & 1 \\ 0 & -18 & -15 & -3 \\ 0 & 9 & 12 & 3 \\ 0 & -2 & -3 & -1 \end{bmatrix} \begin{bmatrix} 1 \\ 1 \\ 1 \\ 1 \end{bmatrix} = \begin{bmatrix} 4 \\ -6 \\ 4 \\ -1 \end{bmatrix}$$

This is, of course, the well-known 4-point prediction algorithm:

$$x(t + T) = 4x(t) - 6x(t - T) + 4x(t - 2T) - x(t - 3T)$$

A more detailed exposition and discussion of the method in general will be found in [Fo85a].

Appendix F: Analysis of the phase advance compensator

Although the main text includes emulation tables, it is useful to know how they are derived, and examples for both the z- and δ-operators are given here. We also give details of how the sensitivity factors are derived.

F.1 The z-filter

F.1.1 Transforming to the z-domain

$$H(z) = G\left[\frac{1 + \dfrac{2(z-1)}{T(z+1)}k\tau}{1 + \dfrac{2(z-1)}{T(z+1)}\tau}\right]$$

$$= G\left[\frac{z(T + 2k\tau) + (T - 2k\tau)}{z(T + 2\tau) + (T - 2\tau)}\right]$$

$$= \frac{G\dfrac{(T + 2k\tau)}{(T + 2\tau)} + G\dfrac{(T - 2k\tau)}{(T + 2\tau)}z^{-1}}{1 + \dfrac{(T - 2\tau)}{(T + 2\tau)}z^{-1}}$$

Hence

$$a_0 = G\frac{T + 2k\tau}{T + 2\tau} \qquad a_1 = G\frac{T - 2k\tau}{T + 2\tau} \qquad b_1 = \frac{T - 2\tau}{T + 2\tau}$$

Note that table 5.2 could have been used to determine these.

F.1.2 Reverse-transforming to the s-domain

The reverse bilinear transform uses $z^{-1} = \dfrac{1 - sT/2}{1 + sT/2}$ so

$$H(s) = \frac{a_0 + a_1\dfrac{1 - sT/2}{1 + sT/2}}{1 + b_1\dfrac{1 - sT/2}{1 + sT/2}}$$

which can be re-arranged to give

$$H(s) = \frac{\dfrac{a_0 + a_1}{1 + b_1} + s\dfrac{a_0 - a_1}{1 + b_1}\dfrac{T}{2}}{1 + s\dfrac{1 - b_1}{1 + b_1}\dfrac{T}{2}}$$

Again, table 5.3 could have been used for this.
Hence, expressions for G, K and τ are:

$$G = \frac{a_0 + a_1}{1 + b_1} \qquad k = \frac{a_0 - a_1}{a_0 + a_1} \cdot \frac{1 + b_1}{1 - b_1} \qquad \tau = \frac{1 - b_1}{1 + b_1}\frac{T}{2}$$

F.1.3 Calculation of sensitivity factors

$$s_{G,a_0} = \frac{\partial G}{\partial a_0} \cdot \frac{a_0}{G} = \frac{1}{1 + b_1} \cdot \frac{a_0}{G}$$

Substitute for b_1 and a_0 (using expressions from section F.1.1) to give the sensitivity factors in terms of G, k and τ:

$$s_{G,a_o} = \frac{1}{1 + \dfrac{T - 2\tau}{T + 2\tau}} \cdot \frac{G\dfrac{T + 2k\tau}{T + 2\tau}}{G} = \frac{T + 2k\tau}{2T}$$

s_{G,a_1} is very similar in derivation to s_{G,a_0}

$$s_{G,b_1} = \frac{\partial G}{\partial b_1} \cdot \frac{b_1}{G} = \frac{-(a_0 + a_1)}{(1 + b_1)^2} \cdot \frac{b_1}{G}$$

Substitute for a_0, a_1 and b_1 to give:

$$s_{G,d_1} = -\frac{\left\{\dfrac{2GT}{T+2\tau}\right\}}{\left\{\dfrac{2T}{T+2\tau}\right\}^2} \cdot \frac{\dfrac{T-2\tau}{T+2\tau}}{G} = \frac{2\tau - T}{2T}$$

$$s_{k,a_0} = \frac{\partial k}{\partial a_0} \cdot \frac{a_0}{k} = \left\{\frac{1}{a_0 + a_1} - \frac{a_0 - a_1}{(a_0 + a_1)^2}\right\} \frac{a_0}{k} \cdot \frac{1 + b_1}{1 - b_1}$$

$$= \frac{2a_1}{(a_0 + a_1)^2} \frac{a_0}{k} \frac{1 + b_1}{1 - b_1}$$

Substituting for a_0 and a_1 gives:

$$s_{k,a_0} = \frac{T^2 - 4k^2\tau^2}{4k\tau T}$$

By inspection we can also write down:

$$s_{k,a_1} = -\frac{(T^2 - 4k^2\tau^2)}{4k\tau T}$$

$$s_{k,b_1} = \frac{\partial k}{\partial b_1} \cdot \frac{b_1}{k} = \frac{a_0 - a_1}{a_0 + a_1}\left(\frac{1}{1 - b_1} - \frac{1 + b_1}{(1 - b_1)^2}\right) \frac{b_1}{k}$$

$$= \frac{a_0 - a_1}{a_0 + a_1}\left(\frac{2b_1^2}{(1 - b_1)^2}\right) \frac{1}{k}$$

Substituting this time for all three coefficients will give:

$$s_{k,b_1} = \frac{T^2 - \tau^2}{2T\tau}$$

τ is dependent only upon b_1, which means that both s_{τ,a_0} and s_{τ,a_1} are zero:

$$s_{\tau,b_1} = \frac{\partial \tau}{\partial b_1} \cdot \frac{b_1}{\tau} = \left[\frac{-1}{1 + b_1} - \frac{(1 - b_1)}{(1 + b_1)^2}\right] \cdot \frac{T}{2} \cdot \frac{b_1}{\tau}$$

$$= \frac{2}{(1 + b_1^2)} \cdot \frac{Tb_1}{2\tau}$$

and the usual substitutions give

$$s_{\tau,b_1} = \frac{T^2 - 4\tau^2}{4T\tau}$$

F.2 The δ-filter

F.2.1 Transforming to the δ-domain:

$$H(\delta) = G\left[\frac{1 + \dfrac{2\delta}{T(2+\delta)}k\tau}{1 + \dfrac{2\delta}{T(2+\delta)}\tau}\right]$$

$$= G\left(\frac{\delta(T + 2k\tau) + 2T}{\delta(T + 2\tau) + 2T}\right)$$

$$= \frac{G\dfrac{T + 2k\tau}{T + 2\tau} + G\dfrac{2T}{T + 2\tau}\delta^{-1}}{1 + \dfrac{2T}{T + 2\tau}\delta^{-1}}$$

Comparison with the standard modified canonic form

$$H(\delta) = \frac{p + q\delta^{-1}}{1 + d_1\delta^{-1}}$$

gives

$$p = G\frac{T + 2k\tau}{T + 2\tau} \qquad q = G \qquad d_1 = \frac{2T}{T + 2\tau}$$

(exactly as table 5.4 would have given).

F.2.2 Reverse-transforming to the s domain

The reverse bilinear transform uses $\delta^{-1} = \dfrac{1 - sT/2}{sT}$:

$$H(s) = \frac{p + qd_1 \cdot \dfrac{1 - sT/2}{sT}}{1 + d_1 \cdot \dfrac{1 - sT/2}{sT}}$$

$$= \frac{qd_1 + s(pT - qd_1T/2)}{d_1 + s(T - d_1T/2)}$$

$$= \frac{q + s\left(\frac{p}{d_1} - \frac{q}{2}\right)T}{1 + s\left(\frac{1}{d_1} - \frac{1}{2}\right)T}$$

and therefore $G = q$

$$k = \frac{1}{q} \cdot \frac{2p - d_1q}{2d_1} \cdot \frac{2d_1}{2 - d_1} = \frac{2p - d_1q}{q(2 - d_1)}$$

and $$\tau = \left(\frac{1}{d_1} - \frac{1}{2}\right) T$$

F.2.3 Calculation of sensitivity factors

It is immediately obvious that

$$s_{G,p} = 0 \qquad s_{G,q} = 1 \qquad s_{G,d_1} = 0$$

$$s_{k,p} = \frac{\partial k}{\partial p} \cdot \frac{p}{k} = \frac{2}{q(2 - d_1)} \cdot \frac{p}{k}$$

Substitution for p, q and d_1 from section F.2.1 gives

$$s_{k,q} = \frac{T + 2k\tau}{2k\tau}$$

$$s_{k,q} = \frac{\partial k}{\partial q} \cdot \frac{q}{k} = \frac{-2p}{q^2(2 - d_1)} \cdot \frac{q}{k}$$

which, with the usual substitutions, becomes

$$s_{k,p} = -\frac{(T + 2k\tau)}{2k\tau}$$

$$s_{k,d_1} = \frac{\partial k}{\partial d_1} \cdot \frac{d_1}{k}$$

$$= \left[\frac{-q}{q(2 - d_1)} + \frac{2p - d_1q}{q(2 - d_1)^2}\right] \frac{d_1}{k}$$

After some simplification, and substitution for p, q and d_1, we obtain:

$$s_{k,d_1} = \frac{(k-1)T}{2k\tau}$$

τ is dependent only upon d_1, so:

$$s_{\tau,p} = s_{\tau,q} = 0$$

and

$$s_{\tau,d_1} = \frac{\partial\tau}{\partial d_1} \cdot \frac{d_1}{\tau} = -\frac{T}{d_1^2} \cdot \frac{d_1}{\tau}$$

$$= -\frac{T}{\tau} \cdot \frac{T+2\tau}{T} = -\frac{(T+2\tau)}{\tau}$$

Appendix G: A second-order filter

The C program below simulates a second-order filter in canonical form, allowing quantisation of variables to a specified wordlength.

```
#include <stdio.h>                        /*  I/O procedures  */
#include <math.h>                          /* Maths procedures */
main()
{
   double u=<input_size>, time=0;
   double v0_exact, v1_exact=0, v2_exact=0, y_exact;
   double v0_trunc, v1_trunc=0, v2_trunc=0, y_trunc;
   double a0=<coef_value>, a1=<coef_value>, a2=<coef_value>;
   double b1=<coef_value>, b2=<coef_value>, T=<sample_time>;
   double factor, j=<no_of_bits>;
   int i;

   factor= pow (2,j);              /* Evaluate 2^j */

   while (time <= <final_time>)
   {
      for (i=1;i<=10;i++)     /* to print every 10 samples */
      {
         v0_exact= (u - b1*v1_exact - b2*v2_exact);
         v0_trunc= floor(factor*(u - b1*v1_trunc - b2*v2_trunc))/factor;
         y_exact = a0*v0_exact + a1*v1_exact + a2*v2_exact;
         y_trunc = a0*v0_trunc + a1*v1_trunc + a2*v2_trunc;
         if (i==1)
            printf("Time= %3.2f Exact= %5.3f Trunc= %5.3f\n",time,y_exact,y_trunc);
         v2_exact = v1_exact;
         v1_exact = v0_exact;
         v2_trunc = v1_trunc;
         v1_trunc = v0_trunc;
         time= time+T;
      }
   }
}
```

Appendix H: Design Structure Diagrams – a brief guide

This appendix gives a summary of the principles and important symbols for Design Structure Diagrams, as detailed in British Standard 6224 (1987).

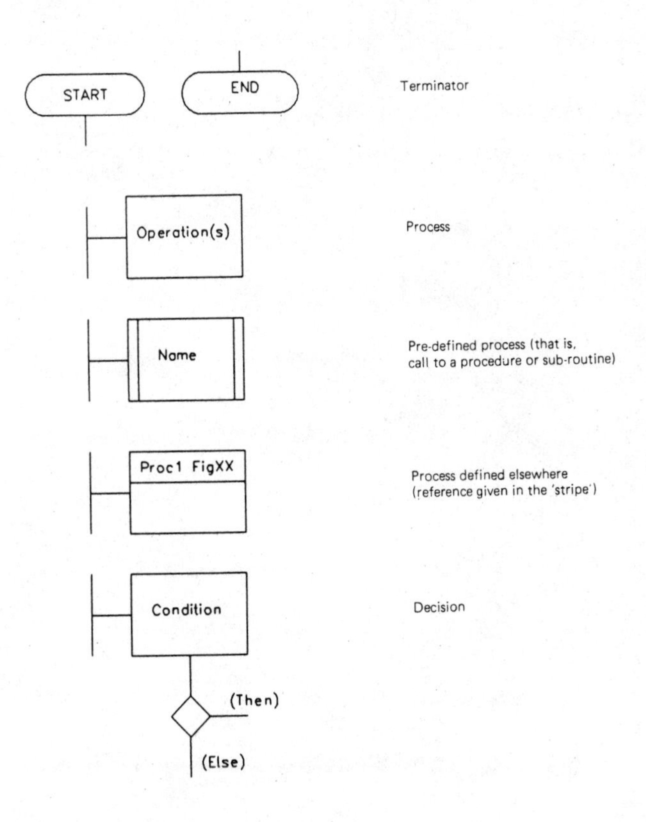

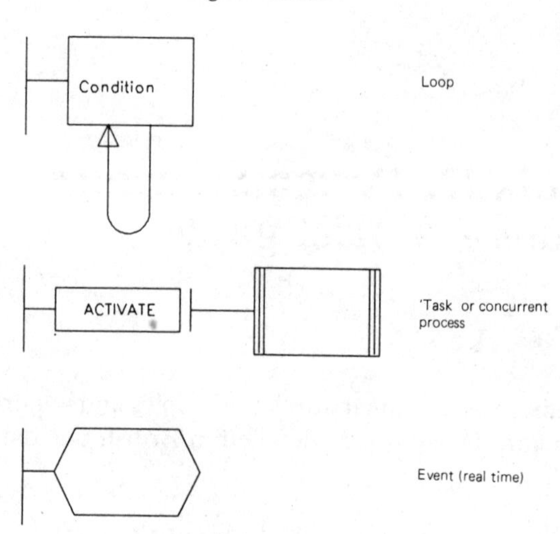

Flow of control: Always turn left at nodes. When activities in left-hand path are complete, fall-back to the previously visited nodes and continue.

References

Ag75 R. C. Agarwal and C. S. Burrus, 'New recursive digital filter structure having very low sensitivity and round-off noise', *IEEE Trans. on Circuits and Systems*, **CAS–22**(12), pp. 921–927, 1975.

As84 K. J. Astrom and B. Wittenmark, *Computer Controlled Systems*, Prentice-Hall, Englewood Cliffs, NJ, 1984.

As89 K. J. Astrom and B. Wittenmark, *Adaptive Control*, Addison-Wesley, London, 1989.

Be88 S. Bennett, *Real Time Computer Control: An Introduction*, Prentice-Hall International (UK), London, 1988.

BS87 British Standards Institution, *BS6224 Design structure diagrams for use in program design and other logic applications*, London, 1987.

Cl83 H. Clarke, '6809 Macros for structured programming', *Micro – The 6502/6809 Journal*, No. 52, pp. 57–63, 1983.

Co81 A. J. Cole, *Macro Processors*, Cambridge University Press, 1981.

Co87 J. E. Cooling, *Real Time Interfacing*, Van Nostrand Reinhold, Wokingham, Berks., 1987.

Fo85a W. Forsythe, 'A new method for the computation of digital filter coefficients', Parts 1 & 2, *Simulation*, **44**(1 & 2), Jan./Feb. 1985.

Fo85b W. Forsythe, K. L. Sherit and A. W. Self, 'Digital compensation: a comparison of methods', *Trans. Inst. Meas. and Control*, **7**(3), pp. 117–126, Apr.–June 1985.

Fo86 W. Forsythe, 'Determination of the truncation error in difference equations', *Applied Mathematical Modelling*, **10**, pp. 367–375, 1986.

Fo89 W. Forsythe, 'Quantization errors in digital filters and their dependence on sampling frequency', *Trans. Inst. Meas. and Control*, **11**(1), pp. 48–56, Jan.–Mar. 1989.

Fr90 C. F. Franklin, J. D. Powell and M. L. Workman, *Digital Control of Dynamic Systems*, Addison-Wesley, London, 1990.

Go85 R. M. Goodall and D. S. Brown, 'High speed digital controllers using an 8 bit microprocessor', *Software and Microsystems*, **4**(5/6), pp. 109–116, 1985.

Go89a R. M. Goodall and D. G. Sutton, 'A low cost solution for high-speed digital control', *Trans. Inst. Meas. and Control*, **11**(1), pp. 9–14, Jan.–Mar. 1989.

Go89b R. M. Goodall, 'Macros – a neglected technique in assembly language programming', *Microprocessors and Microsystems*, **13**(7), pp. 437–444, 1989.

Goo85 G. C. Goodwin, 'Some observations on robust estimation and control', *Proc. 7th IFAC Sympos. Identification and System Parameter Estimation, York (UK)*, 1985.

Ho85 C. H. Houpis and G. B. Lamont, *Digital Control Systems*, McGraw-Hill, International Student Edition, Singapore, 1985.

Ja72 L. B. Jackson, 'On the interaction of roundoff noise and dynamic range in digital filters', in *Digital Signal Processing* (ed. Rabiner and Rader), IEEE Press, New York, 1972.

Ka81 P. Katz, *Digital Control using Microprocessors*, Prentice-Hall International (UK), London, 1981.

Le85 J. R. Leigh, *Applied Digital Control*, Prentice-Hall International (UK), London, 1985.

Li71 B. Liu, 'Effect of finite word length on the accuracy of digital filters – a review', *IEEE Trans. on Circuit Theory*, **CT18**(6), pp. 670–677, 1971.

Mi90 R. H. Middleton and G. C. Goodwin, *Digital Estimation and Control: A Unified Approach*, Prentice-Hall, Englewood Cliffs, NJ, 1990.

Mu77 N. Munro, 'Composite system studies using the connection matrix', *International Journ. Control*, **26**(6), pp. 831–839, 1977.

Ny28 H. Nyquist, 'Certain topics in telegraph transmission theory', *AIEE Trans.*, **47**, pp. 617–644, 1928.

Og87 K. Ogata, *Discrete-time Control Systems*, Prentice-Hall, Englewood Cliffs, NJ, 1987.

Or81 J. M. Ortega and W. G. Poole, *An Introduction to Numerical Methods for Differential Equations*, Pitman, London, 1981.

Ph90 C. L. Phillips and H. T. Nagle, *Digital Control System Analysis and Design*, Prentice-Hall, Englewood Cliffs, NJ, 1990.

Ru90 L. Rundquist, 'Anti-reset windup for PID controllers', *Proc. 11th IFAC Congress, Tallinn, USSR*, **8**, pp. 146–151, 1990.

Sa81 S. V. Salehi, 'Signal flow graph reduction of sampled data systems', *International Journ. Control*, **34**(1), pp. 71–94, 1981.

Se67 M. Sedlar and G. A. Bekey, 'Signal flow graphs of sampled-data systems: a new formulation', *IEEE Trans. Automatic Control*, **AC12**(2), pp. 154–161, 1967.

Sh49 C. E. Shannon, 'Communication in the presence of noise', *Proc. IRE*, **37**, pp. 10–21, 1949.

Sl74 L. I. Slafer and S. M. Levy, 'Sampled-data control system analysis using the Poisson summation rule', *IEEE Trans. Aerospace and Electronic Systems*, **AES–10**(50), pp. 630–635, 1974.

Index

δ-filter 153, 245
δ-form 93, 102, 104
δ-operator 151

Adams–Bashforth algorithms 54
adaptation 114, 191
aliasing 45, 47
analogue, definition of 3
analogue design example 63
analogue *versus* digital 112
analogue-to-digital converter 116, 180
assembly language 191, 204, 209

baseband 47
bilinear transform, definition of 53
bilinear transform with pre-warping 57
bilinear transformation 51, 124, 243

canonic form 121, 153
  modified 155
central processing unit 116
coefficient accuracy 119, 137
coefficient sensitivity 124, 126, 158
computation, precision of 123
connection matrix reduction 38
continuous, definition of 3
controller design 44
co-processor 207
crossover frequency 63
custom integrated circuit 176

data hold 19
data types 205
dead time 25
delta form 93, 102, 104
derivative, use of in emulation 55
design structure diagrams 187, 249
development and testing 181
difference equation 6
digital, definition of 3
digital filters, errors in 89
digital input/output 116, 178
digital signal processor 176, 184, 212
digital-to-analogue converter (DAC) 19, 116, 180
Dirac delta function 29
direct form 7, 92, 99, 103, 104, 120
discrete, definition of 3
discrete convolution 9
discrete frequency response from continuous transfer function 57
discrete transfer function 11

embedded controller 172
emulation 51
  comparison of methods 59
  design example 65
  effectiveness of 51
emulation table 125, 132, 156, 157
encoder 179
error
  algorithmic e_a 89, 96, 109
  coefficient representational, e_c 91, 104, 110
  discretisation, e_a 89
  filter, variation with sampling interval 95
  multiple word truncation, e_r 91, 99
  quantisation, e_q 89, 90
  round-off, e_r 91, 103
  sampler quantisation, e_s 90, 97, 110
  total, e_t 89
  truncation, e_a 89
error coefficients, J_i 96
error mechanisms 88
errors
  numerical evaluation of 109
  quantisation 100

filter 45
filter, antialiasing 48
filter error 88
filter structure 88
Final Value Theorem (FVT) 50

first-order hold 19
floating-point numbers 114, 200, 202
form
  canonic 93, 101, 104
  delta 93, 102, 104
  direct 92, 99, 103, 104
  modified canonic 155
fractional bits 143, 145
frequency, negative and positive 47
frequency response, shaping on gain-phase plane 63, 224
frequency response of digital filter 45
frequency response of digital filter from z-plane 70

$G^*(s)$ 29
$g_p(s)$, $g_p(z)$ 21

hardware 171
high-level language 114, 191, 204

impulse 18
in-circuit emulation 172
Initial Value Theorem (IVT) 50
integer numbers 199
integrator wind-up 148
integrators 146, 165
interlocking 178, 189
interrupt 178, 190, 195

loop stability 88, 106
low-frequency gain 105, 141

m, parameter of modified z-transform 26
macros 209
mapping 13
margin
  gain 107
  phase 63, 107
microcomputer 172
microcontroller 174, 182, 183, 212
microprocessor 173, 182, 183, 210
multiloop controllers 193
multiple precision 198, 202
multiplication 201, 202, 208

Nichols Chart 63, 224
nomenclature, points of 24
notch filter 132, 135, 159, 207
number representation 199
numerical routines 198
Nyquist limit 48, 51, 71

offset binary 199
overflow 138, 139, 141, 148, 161

p, parameter of modified z-transform 26
phase advance compensator 63, 71, 127, 157, 206, 224, 242
PID compensator 149
plant
  discrete model 21
  to model 18
pole and zero mapping 56, 90
pole mapping with Taylor expansion 59
polynomial transformation 8
primary strip 13
processor 173
proportional plus integral (PI) compensator 72, 150, 230
pulse response 12
pulse transfer function 21, 25
PWM output 181

quantisation 89, 168
quantised steady-state value 144

random access memory 116
rationale for choice of system parameters 108
read-only memory 116
root locus, design in z-plane 44
root of sum of squares (rss) 130, 134
rounding 143

sampled data loop 18
  controller design for 44
sampled data systems 5, 14
  analysis of 29
  complex, compensation of 75
  frequency response of 72
sampler
  ideal 28
  impulse 28
sampling frequency, choice of 48, 88, 107
sampling interval 88, 106
sampling rates, multiple 79
sensitivity analysis 125, 131
sensitivity factors 125, 133, 243, 246
sensitivity matrix 128, 131, 133, 158
serial input/output 116
Shannon's Sampling Theorem 54
signal flow graph reduction 38

simulation 146, 248
single board computer 172
software 186
software structure 187
stability, gauging reduction in 107
star notation 28
structure, significance of 92
subroutines 209

time delay, including 25
timer/counter 116, 177, 187
transform table, why use? 24
transport lag 25
transputer 176
trapezoidal rule 5, 52
truncation 143, 203, 205
Tustin's method 5, 51
two's complement numbers 139, 199

$u_s(t)$ 14
underflow 138, 143, 163
unit impulse function 29

variables
  intermediate 117
  internal 117, 138

w-plane, design in 44, 61, 77
w-plane design example 66
w′-plane 62
wordlength
  ADC 89, 104, 119
  coefficient 106, 124, 147, 157
  internal variable 119
  processor 89

$x^*(t)$, $x^*(s)$ 29

z-domain 4
z-plane
  design in 45, 62
  some properties of 69
z-transform, modified 25
z-transform table 14, 216
  use of 17
zero frequency gain 48, 50
zero-order hold (ZOH) 19